# MAPS
# &
# CIVILIZATION

# MAPS
# &
# CIVILIZATION

*Cartography in
Culture and Society*

N ORMAN  J.W.  THROWER

THE UNIVERSITY OF CHICAGO PRESS

CHICAGO AND LONDON

Norman J. W. Thrower is professor in the Department of
Geography at the University of California, Los Angeles.

The University of Chicago Press, Chicago 60637
The University of Chicago Press, Ltd., London
© 1972, 1996 by Norman J. W. Thrower
All rights reserved. Published 1996
Printed in the United States of America
05 04 03 02 01 00 99 98 97 96   5 4 3 2 1

ISBN (cloth): 0-226-79971-9
ISBN (paper): 0-226-79972-7

An earlier edition of this work was published as *Maps and
Man: An Examination of Cartography in Relation to Culture and
Civilization,* Prentice-Hall, 1972.

Library of Congress Cataloging-in-Publication Data

Thrower, Norman Joseph William.
    Maps and civilization: cartography in culture and society/
    Norman J. W. Thrower.
        p.     cm.
    Rev. ed. of: Maps & man. Englewood Cliffs, N. J.: Prentice-
    Hall, 1972.
    Includes bibliographical references and index.
    1. Cartography—History.   I. Thrower, Norman Joseph
    William.
    Maps & man.   II. Title.
    GA201.T47   1996
    526'.09—dc20                                    95-24656
                                                        CIP

*To Page, Anne, and Mary*

*I daylie see many that delight to looke on Mappes but yet*
*for want of skill in Geography, they knowe not with what*
*manner of lines they are traced, nor what those lines do*
*signify nor yet the true use of Mappes.*

Thomas Blundeville
A Brief Description of Universal Mappes and Cardes and Their Use
*(London, 1589)*

# Contents

## SIX
*Cartography in the Scientific Revolution and the Enlightenment*
91

## SEVEN
*Diversification and Development in the Nineteenth Century*
125

## EIGHT
*Modern Cartography: Official and Quasi-Official Maps*
162

## NINE
*Modern Cartography: Private and Institutional Maps*
198

## APPENDIX A
*Selected Map Projections*
237

## APPENDIX B
*Short List of Isograms*
247

## APPENDIX C
*Glossary*
249

*Notes*
259

*Illustration Sources*
297

*Index*
301

# *Preface*

*Maps and Civilization* is a revision of the book *Maps and Man,* which was originally written at the behest of students in my courses on cartography, remote sensing of the environment, and geographical discoveries. It was conceived as a supplementary textbook for courses in the sciences, social sciences, and humanities and was warmly received by the academic community. It was also embraced by the large number of professionals and nonprofessionals interested in the topics it covered. Because of the specifications of the series of which it was then a part, it was less than two hundred pages long.

Since *Maps and Man* appeared in 1972, revolutions as profound as any in the past have occurred in the field of cartography. One of these is the application of the computer to cartographic problems, including the production of animated maps. Similarly, imagery produced through continuous surveillance of the earth by instruments carried onboard satellites has emphasized the continuum between image and map. The potentials of these developments were discussed in the earlier text, but they have progressed so far in the past two decades that this new edition surveying the development of cartography is demanded. Likewise, recent research on the maps of indigenous peoples—a topic considered in the previous edition—as well as on earlier Western cartography has advanced to the point that more attention to these topics is merited.

Fortunately the University of Chicago Press, which commissioned this revision, has allowed a greater number of pages and illustrations than were included in the original. However, to keep the book affordable to students, color is again used only on the cover. Of course, within the compass of a still modest-sized volume, a large subject cannot be treated exhaustively. But references to the rich literature of cartography and related topics—as before, contained in the notes—can lead the reader into many profitable and pleasurable avenues. Generally only articles from the voluminous periodical literature considered crucial are cited, since others are to be found in bibliographies in the specialized books that are referenced. The new title reflects the broader view of the subject that has resulted from a radical revision, expansion, and updating of the original work.

This is a book about maps rather than about mapping. Although there are necessary references to mapmaking, these are included only so that the reader may be informed about the maps that result from the use of particular methods. It is not in any sense a "how-to" work, of which a number of good examples already exist. Those who wish to make maps must, of course, learn by actually creating them. However, one can discover a great deal about maps by studying them; hence the illustrations in this work are most important. For a majority of people, even geographers, a knowledge of maps rather than the principles of mapmaking is needed. It is to provide information on the nature and development of maps and the lure and lore of cartography that this book has been prepared. As Edmond Halley, the English natural philosopher and astronomer who was himself a distinguished cartographer, observed in 1683, by the use of the map certain phenomena ". . . may be better understood, than by any verbal description whatsoever."

In presenting this study, the author would like to take the opportunity to thank his professional colleagues, located in many different centers, for their help. Specific contributions are indicated in the text or in the footnotes, but others—whether acknowledged or not—are greatly appreciated. Those acknowledged include my former doctoral students, especially Patricia Caldwell, Anne Canright, John Estes, John Jensen, Leslie Senger, Judith Tyner, and Ronald Wasowski, and master's students, Gerald Greenberg and Roderick McKenzie (both later Ph.D.'s), Tony Cimolino, Matthew McGrath, and Robert Mullens. A larger number of my other graduate and undergraduate students, though unacknowledged, also contributed. The University of California, and particularly its Los Angeles campus, has provided an academic climate that made the book possible. Several of the diagrams in this edition were created by Chase Langford, Staff Cartographer, Department of Geography, UCLA. Resources of many institutions were utilized, especially those of the UCLA libraries: Research, William Andrews Clark, Powell, Map, and Special Collections. At the UCLA Map Library, Carlos Hagen, Director, and Jon Hargis, Portia Chambliss, and Eric Scott were particularly helpful. Betsy Hedberg, Research Assistant for the project, deserves special praise. Particular thanks are also due to Penelope Kaiserlian, Associate Director of the University of Chicago Press, who proposed this revision. My wife Betty provided the encouragement needed to sustain me as I wrote these thoughts on the subject that has occupied most of my professional life.

The earlier book, *Maps and Man,* was characterized as the first general social and cultural survey of the development of cartography. It is hoped that this expanded volume, *Maps and Civilization,* will enhance that reputation.

ONE

# Introduction: Maps of Preliterate Peoples

As a branch of human endeavor, cartography has a long and interesting history that well reflects the state of cultural activity, as well as the perception of the world, in different periods. Early maps from great civilizations were attempts to depict earth distributions graphically in order to better visualize them; like those of so-called "primitive" peoples, these maps served specific needs. Viewed in its development through time, the map details the changing thought of the human race, and few works seem to be such an excellent indicator of culture and civilization. In the modern world the map performs a number of significant functions as a necessary tool in the comprehension of spatial phenomena; a most efficient device for the storage of information, including three-dimensional data; and a fundamental research tool permitting an understanding of distributions and relationships not otherwise known or imperfectly understood. A knowledge of maps and their contents is not automatic; it must be learned, and it is important for educated people to know about maps even though they may not be called upon to make them. The map is one of a select group of communications media without which, McLuhan has suggested, "the world of modern science and technologies would hardly exist."[1]

Though technological in nature, cartography, like architecture, has attributes of both a scientific and an artistic pursuit, a dichotomy not satisfactorily reconciled in all presentations. Some maps are successful in their display of material but scientifically barren, while in others an important message may be obscured because of the poverty of the representation. An amazing variety of maps exist to serve many different purposes, and it is one of the goals of this book to acquaint the reader with some of these forms. Of course, within the compass of a small work it is possible to give only selected examples of various types of maps, but this selection includes a number of landmark maps in the story of cartography. It seemed better to deal with a limited number of maps in detail

1

than to offer encyclopedic coverage of a larger number without much in-depth discussion. Specific map characteristics will be brought out in reference to particular examples, which are drawn from both earlier and contemporary sources. This work is not a treatise on mapmaking but rather one on map appreciation and map intelligence.[2]

The volume may also be thought of as a sourcebook of cartographic forms or as an anthology of maps, charts, and plans that, like all anthologies, reflects the taste and predilections of the collector. It may also be likened to a book of reproductions of works of art, in the sense that the illustrations—even with the accompanying verbal commentary—cannot really do justice to the originals. In this case, the illustrations are in black and white, many are reduced in scale, and some are merely fragments, reinterpretations, or details. But they will have served their purpose well if people are encouraged by reading this book to look at maps critically, to comprehend their strengths and limitations, to use them more intelligently, and perhaps even to collect them. While no substitute for the map library, this book will ideally lead to the better use of such facilities.[3] Neither is this work a substitute for the rich professional literature of cartography upon which it draws, but it may lead the serious reader to consult additional resources.[4]

Cartography cuts across disciplinary lines to a greater extent than most subjects. No one person or area of study is capable of embracing the whole field, and, like workers in other activities, cartographers are becoming more and more specialized, with the advantages and disadvantages this process inevitably brings. "Nevertheless," Hartshorne has asserted, "workers in other fields commonly concede without question that the geographer is an expert on maps. . . . This is the one technique on which they most often come to him for assistance."[5] Accordingly it is incumbent upon all geographers to understand something about cartography as well as the particular branch of geography in which they specialize. In spite of the notable contributions of a few to cartography, many geographers do not possess enough knowledge of maps to serve as advisors to those in other fields who may consult them. This book is intended to help fill this need as well as to promote the use and enjoyment of maps.[6] It is written especially for the nonspecialist who wishes or needs to know something of maps.

One of the main themes of this book is that the modern map can be well designed, even a thing of beauty and elegance, and that earlier workers in the field had no monopoly on this aspect of cartography. Moreover, the view of this author is that, contrary to the opinion of some, the study of cartography has become increasingly exciting in the last century and a half through the application of modern technology.

In recent years the map as a medium of communication has been enriched by new data and has become capable of conveying its messages
in increasingly interesting ways. At the same time, the visual qualities
have been vastly improved through the development of new techniques,
materials, and processes. However, in the present work, some aspects of
contemporary cartography are treated more briefly than their importance suggests because individual books in the field cover such topics
as map transformations, computer mapping, and gravity models. The
emphasis here will be on landmarks of geocartography.

Of course mapping is not confined to the representation of the
earth; other phenomena such as the human brain have been mapped.
The principles and methods of cartography have a universality that
makes them applicable to the mapping of extraterrestrial, as well as terrestrial, bodies. In particular, lunar mapping, which is not a new activity,
will receive some attention here. But the main emphasis will be on what
is called *geocartography*. This term should come into greater use as we
receive and process more detailed information about bodies other than
the earth. As space technology develops, it may be desirable to distinguish between extraterrestrial and terrestrial mapping as we now distinguish between astronomy and geography.

Previously there was general agreement on the meaning of the geographical term *map*, but because of a wider acceptance of images under
this rubric, the borderland between picture and map has recently become blurred. This will be pointed out subsequently in relation to modern imagery. According to the definition used by earlier peoples—and
still used by purists—a map is a representation of all or a part of the
earth, drawn to scale, usually on a plane surface.[7] A wide variety of materials has been used in cartography, including stone, wood, metal, parchment, cloth, paper, and film. The words *map* and *chart* appear to derive
from materials: the Latin word *carta* denotes a formal document on
parchment or paper, and *mappa* indicates cloth. In geography today, the
term *chart* is most often applied to maps of the sea and coasts, or at least
to maps used by sailors and aviators. *Map* is a broader term in modern
usage, referring more particularly to a representation of land, while *plan*
is a representation of a small area from above.

To illustrate some of the foregoing ideas, let us examine the cartographic works of peoples in prehistoric times and of preliterate—so-
called "primitive," "native," and indigenous (non-Western)—societies of
a much more recent date.[8] That such groups engage in mapping attests
to the basic importance of cartography to humankind. What has been
called the oldest known plan of an inhabited site is the Bedolina map
from northern Italy, ca. 2000–1500 B.C. (fig. 1.1).[9] Analysis of this rock

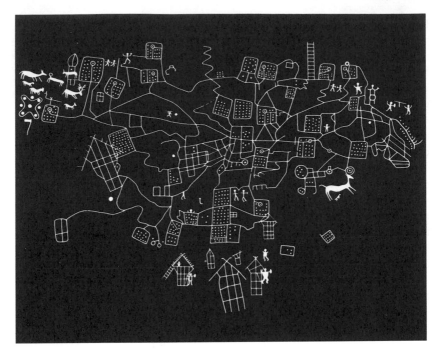

*Figure 1.1.* The Bedolina petroglyph, an example of prehistoric mapping.

carving suggests that it was "engraved" in different stages with pictorial features—human figures, animals, and houses in side view—added later, possibly in the Iron Age, to a Bronze Age plan. Still under debate are the meanings of its more abstract symbols: rectangles filled with regularly spaced points (fields bounded by stone walls?); irregular, single connecting lines (streams and irrigation channels?); and small circles (wells?) with single central points. Whatever the interpretation, the original petroglyph appears to be a detailed plan of an actual area and to represent a "progression" from symbolic to pictorial form. Many other terrestrial and celestial "map" designs in rock art from different parts of the world, whether known or yet to be discovered, would repay careful study by archaeologists, anthropologists, and ethnographers as well as by historians and geographers.

Coming closer to our time, the Pacific Islanders, plotting their courses from island to island, spread over a very large area before Europeans reached the largest of oceans. Again, disciplines beyond history and geography are needed to unravel this story. Among the various homing or way-finding devices used by these interisland navigators are the stars, signs in the seas, land-indicating birds, and "charts." Stick charts from the Marshall Islands illustrate what might be called, for want of a better term, "native" cartography. These charts are generally made of narrow strips of the center ribs of palm fronds lashed together with cord made from locally grown fiber plants. The arrangement of the sticks indicates the pattern of swells or wave masses caused by winds, rather than of currents, as was formerly thought to be the case. The positions of islands are marked approximately by shells (often cowries) or coral. The charts vary considerably in size but are usually between eighteen and twenty-four inches square.[10]

The method for using these charts was elicited from natives of the islands with considerable difficulty because their navigational methods were closely guarded secrets. Distances between the various islands of the Marshall group are not great, but because they are low atolls, the islands can only be seen from a few miles away from an outrigger canoe. To locate an island that is not visible, the native navigator observes the relationship between the main waves, driven by the trade winds, and the secondary waves (reflecting or converging) resulting from the presence of an island. If a certain angle exists between the two sets of waves, a choppy interference pattern is established. When such a zone is reached, the canoe is placed parallel to this pattern with the prow in the direction of the waves of greater amplitude, which give a landward indication. These often complex wave patterns can be illustrated on the stick charts, which may be carried on the canoe. In addition, the naviga-tors lie down in their craft to feel the effect of the waves.

Three major types of charts are found in the Marshall Islands, namely *rebbelib, meddo,* and *mattang.* The *rebbelib* (fig. 1.2) is a chart of a large part of the Marshall group, which consists of about thirty atolls and single islands over a distance of approximately six hundred sea miles northwest–southeast and about half that distance northeast–southwest. Although the spatial relationships between the islands are only approxi-mated on the stick charts, these locations can be recognized by referring to modern navigational charts of the area. The *meddo* is a sectional chart of part of the island group; it may be one of a series of charts, and its scale allows more detail to be shown than is possible on the *rebbelib.* Un-like the others, the third type of Marshallese stick chart, the *mattang,* is not carried on canoes but is used for instructional purposes. A *mattang* is a highly conventionalized, often symmetrical, chart that does not nec-essarily show an actual geographical location. It provides a summary of information about wave patterns that might have wide application, al-though a full understanding of its characteristics may be possessed only by its maker.

In the Marshallese stick charts we can see ingenious, independent, and spontaneous solutions to various cartographic problems. The materials from which they are made (palm fronds, shells, and so on) are available within the limited material resource base of the Islanders, reminding us that not all maps are documents printed on paper. The stick charts illustrate spatial phenomena of infinite importance to the native interisland navigator but of little significance to most other people. *Rebbelib* and *meddo* charts indicate the need for maps of different scales: the *rebbelib* shows a broad area and uses a small scale, while the *meddo* illustrates a more restricted locality on a larger scale. In the use of the *mattang*, we recognize the necessity of learning to read charts or maps to understand the relationship between cartographic conventions and reality. Furthermore, the desire to conceal a body of geographical information and its cartographic expression, as exemplified by the reluctance

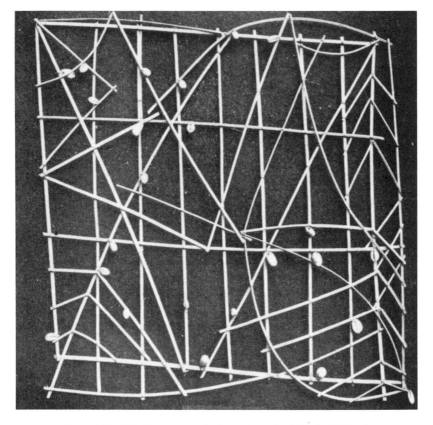

*Figure 1.2. Rebbelib,* or general chart, from the Marshall Islands.

of the Marshall Islanders to share their techniques, is a recurring and continuing theme in the history of mapmaking.

The Marshall Islanders were by no means unique, and when the Europeans came to other areas new to them they usually encountered populations with a superb knowledge of their local regions. Thus when Christopher Columbus arrived at Guanahani (which he renamed San Salvador) in the Bahamas in October 1492, he learned by signs from the inhabitants that there was a larger island to the south: Cuba. Later, when the Spanish reached the Aztec civilization of Central America, they found a well-developed cartography of the area. Hernán Cortés reported that in 1520 he received a chart from Montezuma that showed a large part of the Mexican coast on which were delineated estuaries, rivers, and bays. Such maps were sometimes taken to Europe and information from them used in the compilation of published maps. For example, a map of Mexico City with an inset of the coast of Mexico that was printed in Germany in 1524 apparently owes a good deal to indigenous maps of the area.[11] Such maps were different from anything the Europeans had seen before, but most of these manuscripts have perished and their contribution was seldom acknowledged.

However, a map from Mexico that shows little European influence is the frontispiece of the *Codex Mendoza,* now at the Bodleian Library at Oxford. This work was commissioned by Viceroy Antonio de Mendoza, ca. 1547, although the employment of this name dates only from the late eighteenth century. Mendoza was interested in collecting indigenous cultural information on New Spain for transmission to Europe, where the *Codex* was acquired by Richard Hakluyt from André Thevet, whose name appears on the manuscript. Details of the work were published from the seventeenth century on. An edition using modern methods of color reproduction appeared in 1938 and a new edition in 1992 (fig. 1.3).[12]

The map depicts the founding of Mexico City (Tenochtitlán) with an eagle, the Aztec symbol for the sun, perched on a cactus in the center and the seal of the city below. The rectangle with diagonals is a stylized plan of the settlement and its waterways, the home of some 150,000 people at the time of the conquest. Attempts have been made to interpret the plan and the pictorial symbols in its four divisions, but much of this is speculation. The figures within the quadrants are believed to be the ten founders of the city and the structure at the top center the one-hundred-foot-high temple that greatly impressed the conquistadores. To the right center is a rack for the skulls of sacrificial victims. Below the rectangle are idealized scenes from the Aztec conquests, and the border is a continuous calendric count of fifty-one years, each compartment

*Figure 1.3.* Frontispiece of the *Codex Mendoza,* a manuscript map of Mexico City as conceived by the Aztecs.

representing one year in groups of thirteen with repeating symbols. The Aztecs devised a very sophisticated calendar, and the map is not a snapshot in time but a representation of events that occurred over a number of years in "what was to become one of the greatest pre-Columbian cities ever to flourish in the Western Hemisphere."[13] To the Aztecs, the capital was the center of the cosmos, with the surrounding canals representing the four cardinal directions. Other Aztec maps of more rural areas—such as that of the Tepetlaoztoc district—depict hills, roads, streams, forests, and pyramids with place-names in hieroglyphics and are thus transitional between the cartography of preliterate and literate peoples.

Some study has been devoted to indigenous Mexican maps and those of the Inuit (Eskimo) peoples of the north. But until quite recently maps of the Indians of what is now the United States (Amerindians) had received much less attention. This situation is changing with the establishment of research programs devoted to Amerindian mapping, a distinguished lecture series on the topic, and an exhibition with a number of examples.[14] The analysis of spatial information provided by non-Europeans is problematic because of the use of caricature and exaggeration as well as uncertainty about what the local population might consider important to map.

However, a manuscript map on paper presented by the Iowa Chief Non chi ning ga at a council held in Washington, D.C., in 1837 is recognizably representational (fig. 1.4).[15] It was drawn in ink, possibly with a stick or a finger, and it depicts a large area of the Upper Mississippi and Missouri river drainage systems. Comparison of this delineation with the same hydrography on modern maps shows it to be remarkably comprehensible, although generalized. The present city of St. Louis would be near the confluence of the two largest rivers near the bottom of the map, and major tributaries such as the Platte and the Wisconsin Rivers are identifiable.

Understandably, rivers are a common feature of maps by Amerindians, but tracks and habitations are also represented. So, too, are cosmological ideas and celestial bodies, as in the maps of other indigenous peoples. Rock, tree bark, and animal skins are among the materials used for such native maps, which are painted with natural dyes or scratched on the surface. Many of these works, such as those by the Aborigines of Australia, have considerable artistic merit and are of cultural and religious significance to their makers. An example of the latter is provided by figure 1.5 from the Yolngu community of northeastern Arnhem Land. This bark painting can be interpreted as a map, and it relates to a specific coastal area, Biranybirany. The saltwater crocodile, an ancestral

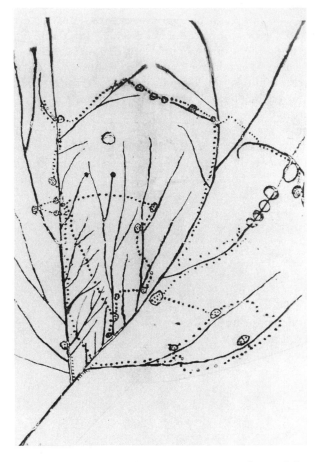

*Figure 1.4.* An Amerindian manuscript map of part of the
Mississippi–Missouri drainage system.

figure, has its feet on ground on which there are no particular terrestrial
boundaries. The Yolngu people move from place to place and are at
home in the landscape; the mouth of the river is where the tail joins the
body. In order to read the map it is necessary to know songs and dances
of the creation of the Ancestral Being and its relatives. The various parts
of the crocodile are named, and children learn the shape of the land
from the bark map.

Thus various groups of preliterate peoples whose very existence depends on knowledge of particular areas make maps,[16] but, as we have
seen, they are not restricted to mundane themes.[17]

*Figure 1.5.* A bark painting from the Yolngu community of northeastern Arnhem Land, Australia, showing the Biranybirany area.

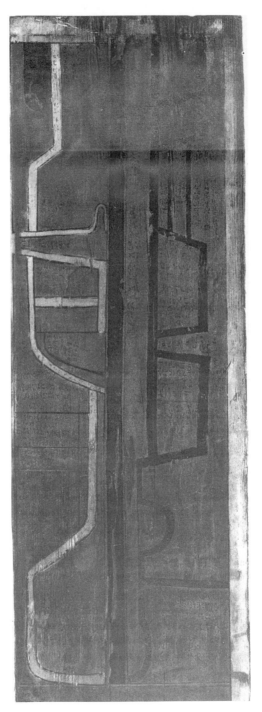

*Figure 2.1.* Early Egyptian map from a wooden sarcophagus.

# TWO

## Maps of Classical Antiquity

We have noted that early preliterate peoples—as well as those of our own time, or close to our time—have used widely different means to express themselves cartographically and that their maps are marked by variety in objective, symbolism, scale, and materials. Similarly, the cartography of literate peoples in antiquity shows remarkable variety in form and function. Only a small fraction of the maps produced in earlier ages has survived, but in some instances we know of lost works through written records. The loss of many early maps, charts, plans, and globes can be attributed to the materials used for their construction, which often militated against their preservation. Thus, valuable metal was melted down and parchment scraped to be used for some other purpose. Alternatively, less durable materials quickly deteriorated, especially when taken to a different climate, or were destroyed by war, fire, or other means. The destruction of maps is a continuing problem, especially because the information they contain may go out of date so quickly that they are treated as ephemera (especially in the computer age) or because they include data of strategic value and therefore cannot be disseminated.

Maps, charts, and plans—both celestial and terrestrial—as well as cosmological schema are part of the record of early civilizations. For example, from Egypt we have a detailed plan of a garden on wood surfaced with plaster, ca. 1500 B.C.; a cosmological map of Egypt carved in stone, ca. 350 B.C.; a zodiacal map also carved in stone, ca. 100 B.C.; and a map of a gold mine in Nubia from the Ramesside Period—part of the so-called Turin papyrus—among other "cartofacts."[1]

In addition to the above are architectural plans of tombs in a variety of media and maps on coffin bases. Several such maps, which predate the map of the gold mine by centuries, were collected from Dar el Bersha in Middle Egypt. They are "passports" to the world beyond, as indicated by a text, *The Book of the Two Ways,* that accompanies the example illustrated as figure 2.1.[2] The "two ways" are represented by light coloring for the day journey and dark for the night journey. The painter used a bluish gray color to represent water (the Nile and a surrounding ocean), with a background of yellow indicating the sands of the desert and, indeed, what can be taken as a continental land mass. This highly stylized rendering may be considered a very early example of a *theoretical map* or model, a cartographic form that persists in spite of the admonitions of some that geographers should address themselves only to the real world. The Egyptians also used realistic pictorial devices, rather than the diagrammatic symbols shown in figure 2.1, in their cartography. It should be mentioned that they are credited with inventing geometry in response to the need for recurrent property surveys as the flooding of the Nile erased boundary markers.[3] While it is believed that cadastral or property maps were made by these people, the only known survivors of this type of map are plans of various buildings and the map of the Nubian gold mines. The production of cadastral maps of more than local areas—of great value to administrators for taxation and other purposes—would be a logical development. But in the absence of survivors from Egypt, we may look to other civilizations for examples of this cartographic genre.

Some of the earliest maps from a civilization that we know about come from Mesopotamia. To illustrate the diversity of scale and purpose in the cartography of this area, we reproduce three examples. Although these maps are different in several particulars, they are alike in that all of them were drawn with cuneiform characters and stylized symbols impressed or scratched on clay tablets, a method that placed great restrictions on the cartographer because it frequently involved using a series of straight lines to approximate curves. This makes some of these early maps strangely reminiscent of some early computer cartography, in which the constraints of a machine often forced the approximation of curving lines with straight line segments (see fig. 9.12 in chap. 9). We will consider the three Mesopotamian maps in terms of scale, dealing first with that of the largest scale, which depicts a small area in the Nippur district.[4] This fragment (fig. 2.2) shows canals of varying widths, a city wall with gates and a moat, houses and their openings, a park, and so on. Features are identified by name on this plan, which dates from ca. 1500 B.C., and the claim is made that it is correct in scale.

A Mesopotamian map of intermediate scale—though of such small dimensions that it can be held in the palm of the hand—is the well-known Akkadian map found at Nuzi and dated ca. 2300 B.C. (fig. 2.3). Sometimes described as the oldest map in the world, this map is oriented with east at the top, and certain features can be clearly identified. These include water courses, settlements, and mountains. The latter are shown by scalelike symbols at the top and bottom of the map, an atypical form of representation compared to the plan format, which is used for other features and characterizes Mesopotamian cartography in general.

The third item in this series (fig. 2.4) is a world map with an Assyrocentric view and, naturally, is of smaller scale than the other examples. It shows a round—though presumably flat—earth with Babylon in the

*Figure 2.2.* Early Mesopotamian city plan on a clay tablet.

*Figure 2.3.* Early map of Mesopotamia on a clay tablet.

center. The Euphrates flows from its source in the Armenian mountains (in the north) to the Persian Gulf, where it joins an encircling sea. Indeed, the purpose of the map seems to be to show the relationship between the "Earthly Ocean," represented by a circle, and the "Seven Islands" (distant places), illustrated by triangles, only one of which is entirely intact on the tablet. In addition to Babylon, several other cities are shown by circles and some are named, as is the swamp at the lower left of the diagram. The astrological and religious significance of the map is indicated in the text of the tablet. Concerning this we need only say that other civilizations, understandably, have taken an egocentric view of the world; the long-held geocentric theory may be the ultimate expression of such an idea applied to the universe. Looking forward, we may note here that the separation between the terrestrial and the celestial spheres was an important theme in the Middle Ages in Europe, as was the concept of a circumfluent ocean. This last idea was also current

*Figure 2.4.* Early Mesopotamian world map on a clay tablet.

among certain Greek philosophers. In considering the circle in cartography, it should be recalled that the sexagesimal system of dividing this figure, which is the usual method employed in mapping to this day, came to us from Babylon by way of Greece.

Maps that are known to us only through descriptions or references in the literature (having either perished or disappeared) are a problem to historians of cartography. In some instances, general works include later reconstructions in the correct chronological order of the original maps, and this practice can be amply justified when a particular area or civilization forms the main basis of discussion.[5] In the present volume, however, the principal emphasis is upon map representation, so it is difficult to take this approach. All reconstructions are, to a greater or lesser degree, the product of the compiler and the technology of the times. Therefore, map reconstructions, in contrast to explanatory diagrams, will be used here only to illustrate the cartography of the period in which the particular map was made. Nevertheless, reconstructions of maps that are known to have existed, and that were made a long time after the missing originals, can be of great interest and utility to scholars. The possibilities include those for which specific information is available to the compiler and even those that are described or merely referred to in the literature. Of a different order, but also of interest, are those maps made in comparatively recent times that are designed to illustrate the geographical ideas of a particular person or group in the past but are suggested by no known maps.[6] Modern scholars have often used reconstructions of maps to illustrate the cartography of antiquity. Although not used as illustrations in this section of the work, reconstructions of— or at least maps that owe their inspiration to—those of the Greeks and Romans will be used in chapters 4 and 5, which treat the periods in which these maps were actually made.

The lack of direct cartographic evidence has made consideration of early Greek mapping speculative. However, a series of Greek coins dating from ca. 330 B.C. was recently found, and one of the images is claimed to be "the earliest Greek map to come down to us in any form and the first physical relief map known."[7] As shown by figure 2.3, the latter part of this claim cannot be substantiated, but the relief features of the coin, an Ionian tetradrachm, are viewed from above rather than in profile, which represents a considerable advance over the earlier example as well as over later ones. The early Greeks were philosophers more than experimenters, and there is continuing debate among contemporary scholars about their hypotheses; we will deal briefly with some of these as they relate to cartography and geography. There was rarely a clear distinction between these two disciplines in antiquity, and a number of workers who were designated geographers were, in fact, cartographers. However, there was a sharp dichotomy in ancient Greece between land measurers (geometers), who were employed to delineate small areas, and philosophers, who speculated on the nature and form of the entire earth. The idea of the earth as a slab eventually gave way to that of a drum- or pillar-shaped world and later to that of a circular form. As early as ca. 550 B.C., the Ionian Anaximander of Miletus drew a world map that was improved upon fifty years later by Hecataeus, also from Miletus. He divided the world into two parts: Europe, and Asia with Africa (Libya). The Mediterranean was in the center of a world island composed of the lands bordering this sea, the whole being surrounded by a circumfluent ocean (ocean-stream). It is assumed that the gnomonic projection (see appendix A) was developed by the early Ionian philosophers, possibly Thales, but that it was used only for astronomical purposes at this time. There was much discussion among the Ancients concerning the major divisions of land area, from which arose the concept of the three continents of the Old World. Through his travel writings Herodotus (active 440–425 B.C.), though not considered a geographer, did much to enlarge contemporary knowledge of Asia. The travels toward northern India of Alexander the Great (d. 323 B.C.), who used *bematists,* or pacers, to produce estimates of distances traveled, further enlarged the world known to the Greeks. Democritus (d. ca. 370 B.C.) considered the inhabited world *(oikoumene)* to be one and a half times as long (east–west) as broad (north–south), a proportion accepted by Dicaearchus (d. 296 B.C.).

Meanwhile, an idea fundamental to later progress in cartography, the spherical form of the earth (which apparently had its beginnings among the Pythagoreans), was gaining currency through the work of Plato (d. ca. 347 B.C.) and his followers, including Aristotle (d. 322 B.C.). Although espousing a geocentric universe, Aristotle proposed a division of the globe into five climatic zones, with the inhabited earth bounded by a protoarctic circle in the north and a tropic circle in the south. These lines were later to be formalized as 66½ and 23½ degrees north latitude, respectively. It was further postulated that there would be a similar habitable (if not inhabited) zone in the Southern Hemisphere, bounded by a tropic line and an antarctic circle (23½ and 66½ degrees south). Eventually these solstistial lines were expressed cartographically as small circles of latitude and along with the great circle of the equator (thought by the Greeks to be the center of an uninhabitable zone by reason of heat) have become a feature of globes and world maps up to the present time. Of course some of these relationships were understood by earlier civilized peoples—as well as today by contemporary groups such as the

Eskimos, to whom they are of infinite importance—but as in so many instances it was the Greeks who systematized this body of knowledge. The Aristotelians also enlarged the theoretical inhabited world considerably, to a length twice as great as its breadth.

A scholar with vision large enough to put this information into a logical framework was needed, and such appeared in the person of Eratosthenes (276–196 B.C.). Head of the library at Alexandria from 240 B.C. until his death, Eratosthenes was known as *beta* to his contemporaries because they considered him second to other Greek savants in all his varied academic pursuits. More critical of these accomplishments was Strabo (ca. 63 B.C.–A.D. 24), to whom we are indebted for much of our knowledge of geography in antiquity, including the work of Eratosthenes. Later workers have a higher opinion of Eratosthenes, regarding him as "the parent of scientific geography"[8] and at least "worthy of *alpha*" in that subject, particularly for his remarkable measurement of the circumference of the earth.[9] Once the idea of a spherical earth was accepted, the measurement of this body was a logical step. Eratosthenes was not the first to compute a figure for the circumference of the earth; this distinction may belong to Eudoxus of Cnidus (d. ca. 355 B.C.), who estimated its measurement at 400,000 stades and who also made a celestial globe, no longer extant.[10] A figure of 300,000 stades is credited to Dicaearchus, a student of Aristotle. A similar figure was proposed by Aristarchus of Samos (d. 230 B.C.), who has been called the "Copernicus of Antiquity" because of his early espousal of a heliocentric rather than a geocentric view of the universe. Perhaps more properly, Copernicus should be called the "Aristarchus of the Renaissance."

Both the method and the accuracy of Eratosthenes' well-known measurement of the earth have evoked the admiration of later workers, and this calculation is regarded as one of the great achievements of Greek science. Eratosthenes observed that at midday during the summer solstice, the rays of the sun fell directly over Syene (Aswan) and that the vertical rod of the sundial (*gnomon*, or style) would not cast a shadow. At the same time of the day and year, the shadow cast by a gnomon at Alexandria, to the north of Syene, was measured by Eratosthenes as 1/50 of a proper circle (fig. 2.5). He assumed that Syene (S) and Alexandria (A) lie under the same meridian circle; that rays ($R^1$ and $R^2$) sent down from the sun are parallel; that straight lines falling on parallel lines make alternating angles equal; and that arcs subtended by equal angles are similar (angle ACB is equal to angle SZA). He accepted a figure of 5,000 stades for the distance from Syene to Alexandria, which, according to his previous reasoning, was 1/50 of the circumference of the earth. Thus 5,000 stades times 50 equaled 250,000 stades for the whole circum-

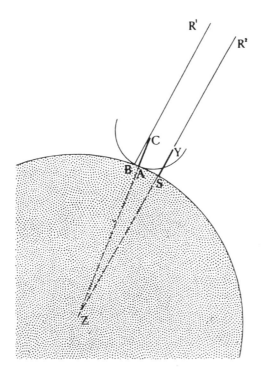

*Figure 2.5* Diagram of Eratosthenes' measurement
of the circumference of the earth.

ference of an earth assumed to be perfectly spherical. There has been much debate about the length of the stade, and we know that Syene is approximately 3 degrees east of Alexandria and some thirty-seven miles north of the tropic. Nevertheless, it is now thought that Eratosthenes' measurement may have been within two hundred miles of the correct figure of the circumference of the earth. To make the measure of the great circle divisible within the sexagesimal system, Eratosthenes later extended the value from 250,000 to 252,000 stades.

Like his Greek predecessors, Eratosthenes also attempted to divide the earth in a meaningful way. In this, too, he followed Dicaearchus, who had separated the known inhabited world into northern and southern and eastern and western parts, with one line passing through the Pillars of Hercules (Strait of Gibraltar) eastward to what is now Iran (the *diaphragma*) and another perpendicular to it passing through Rhodes. Eratosthenes accepted this proto parallel-and-meridional schema but went further and divided the inhabitable world into unequal, straight-sided, geometrical figures compatible with the shapes of different countries.

Although this was not successful, some have seen it as the forerunner of the development of true map projections. Other accomplishments of Eratosthenes affecting cartography were his measurement of various distances on the earth's surface, including the length of the Mediterranean (the best for the following thirteen centuries), and his considerable additions to the world map, especially in southern Asia and northern Europe, made possible through information provided by travelers. Eratosthenes also argued for a predominantly water-covered earth in contrast to others, such as Crates, who proposed a largely terrestrial sphere (fig. 2.6).

Crates of Mellos (active during the second century B.C.) constructed a large globe on which were delineated four approximately symmetrical continents: two in the northern hemisphere and two in the southern, separated from one another by relatively narrow, circumfluent bodies of water (later Oceanus). According to this design, then, there were four large landmasses, three in addition to the known inhabited world. As we shall see, this idea persisted for centuries. A contemporary of Crates, the famous astronomer Hipparchus of Nicaea, is considered to be the true originator of map projections. He devised a systematic, imaginary grid of equally spaced parallels *(climata)* and meridians crossing one another at right angles. Although a few points were astronomically fixed, positions of places on the earth were estimated, largely because of the diffi-

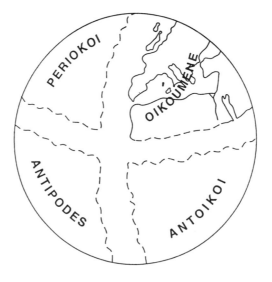

*Figure 2.6.* Crates' globe.

culty of measuring longitude. Hipparchus, who used 360 degrees of 700 stades each for the earth's circumference, also insisted on the accurate location of places according to latitude and longitude, as determined by astronomical observation. Once a systematic grid was adopted for the earth, the serious study of map projections was possible. In fact, the stereographic and orthographic projections, which later became popular for earth maps but in his time may have been used only in astronomy, are attributed to Hipparchus. He is also credited with the invention of the astrolabe (star measurer). In the first century A.D., Marinos (Marinus) of Tyre, like Eratosthenes and Hipparchus before him, attempted to enrich the world map by incorporating information from new sources. Marinos also devised a simple rectangular (plane) chart based on the latitude of Rhodes. These contributions influenced Ptolemy, a later critic of Marinos, to greater accomplishments in cartography.

Claudius Ptolemy (Klaudios Ptolemaios), who flourished in the second century A.D., was librarian at Alexandria, the position held some four centuries earlier by Eratosthenes. Like his predecessor, Ptolemy made giant strides in various phases of cartography that were not materially improved upon for many centuries after his death. Ptolemy did not use Eratosthenes' measurement of the earth but rather employed the smaller measure of the Greek astronomer Poseidonius (186–135 B.C.). This "corrected" figure of the circumference of the earth—some 180,000 stades, or roughly three-quarters of the actual distance—had also been used by Marinos and others. Because of its adoption by Ptolemy, whose authority as both astronomer and cartographer was not seriously challenged for fourteen centuries following, the error was perpetuated. Nevertheless, Ptolemy's specific contributions to cartography were of the greatest importance. They are contained, along with material from other scholars, in his guide to making maps, a work now known as the *Geographia*, or simply the *Geography*.[11] Ptolemy's *Geographia* included instructions for making map projections of the world (coniclike, with straight, radiating meridians and concentric parallels; and one with curved meridians and curved parallels); suggestions for breaking down the world map into larger-scale sectional maps (in some editions there are twelve of these for Asia, ten for Europe, and four for Africa—twenty-six in all—while others propose a much larger number of regional maps); and a list of coordinates for some eight thousand places. For the last of these, two different systems were employed: latitude and longitude in degrees; and latitude according to the length of the longest day with longitude in time (one hour equals 15 degrees) from a prime meridian. Ultimately Ptolemy's prime meridian (0 degrees longitude) passed through the Fortunate (Canary) Islands, and his map extended

180 degrees eastward to China (see figs. 5.2 and 5.3 in chap. 5). Although, like Hipparchus, Ptolemy argued for the astronomical determination of earth locations, in fact most of those contained in the *Geographia* were supplied by travelers and were based on dead reckoning (position from courses sailed and distances made on each course). Also, like Hipparchus, who is often regarded as the greatest astronomer of antiquity, Ptolemy espoused the geocentric theory of the universe. The influence of these two scholars perpetuated the error until the (near) posthumous publications of Copernicus (d. 1543).

We are not sure that Ptolemy actually made maps, and in any case he was greatly indebted to others, including Eratosthenes and Marinos, for both concepts and data. Ptolemy's work has come down to us through later copies preserved in the Byzantine Empire. From these manuscripts, some of which contain maps, it has been possible for scholars to reconstruct knowledge of the world in the centuries immediately following the birth of Christ. As we shall see later, Ptolemy was the ultimate authority on cartography (and astronomy) at the beginning of the great European overseas geographical discoveries of the Renaissance. Perhaps it is incorrect to attribute all these developments specifically to Ptolemy and better to think of a Ptolemaic corpus—similar to the Hippocratic tradition in medicine—to which a number of workers contributed.

Caesar controlled the Alexandria of Ptolemy, who, like Strabo before him and later Greek scholars, labored for Roman masters. Through such means, the Romans became the heirs of the geographical knowledge of the Greeks, which, as we have seen, included the idea of a spherical earth; measurements of the circumference of the earth; irregular and regular division of the sphere (coordinate systems); map projections; maps of different scales; and a world map that embraced large parts of Europe, Africa, and Asia and that, understandably, was progressively less accurate as distance from the Mediterranean increased. From the available evidence, the Romans appear to have been eminently practical in their own cartographic work, being concerned with maps to assist in the military, administrative, and other concerns of the empire. It is possible that this pragmatic approach has been overstressed by scholars, but the evidence seems to bear it out. Our knowledge of Roman cartography comes from a few surviving examples and through references in literature that indicate that, in addition to archival copies on fabric, stone, or metal, duplicates were publicly displayed. One such world map inspired by Augustus Caesar (27 B.C.–A.D. 14) was made by his son-in-law Marcus Vipsanius Agrippa and completed by others. There is much dispute about this map, but it is credited with being the first to actually

*Figure 2.7.* Part of the Orange Cadaster carved in marble.

show the inhabited world divided into three continents, Europe, Asia, and Africa (Libya). It was clearly didactic in purpose and thus is a forerunner of the familiar classroom wall map.

We know of Agrippa's map only through references, but we are on much safer ground when we consider Roman cadastral (property) surveys, of which more tangible evidence exists. Anticipating the United States Public Land Survey by nearly two thousand years, rectilinear surveys with numbering systems were laid out in vast areas of the Roman Empire, from North Africa to Britain. The method used is called *centuriation* from the internal division of hundreds, the right-angled corners of which were laid out with an instrument known as a *groma*.[12] An image of such a Roman surveying instrument is found on a tombstone from the first century B.C. We also have fragments of maps *formae* carved in marble showing an area surveyed into hundreds near Orange (Arausio) in the south of France (fig. 2.7).[13] This fragment shows the grid pattern of property boundaries, with land ownership information inscribed. Such mapping began on the reclaimed lands of the Campagnia outside Rome, where an archive was established ca. 170 B.C. Not all Roman centuriation displays consistent orientation, as illustrated by drawings in the *Corpus Agrimensorum,* a later (ca. A.D. 500) compilation of Roman surveyors' manuals, and by evidence in the European and North African landscape today.

Agrippa's small-scale world map stands in contrast to the large scale of cadastral surveys. Other examples of Roman (or Roman-inspired) cartography exist, including town plans, a mosaic of hydrological features at intermediate scale, and itineraries. One of the last of these is a design drawn on a parchment covering a Roman soldier's shield (ca. A.D. 260) showing a road along the shore of the Black Sea (Pontus Euxinus) with place-names and mileages. However, the best-known example of this type of map is the *Tabula Peutingeriana,* which derives from the fourth century but exists only as a twelfth- or early-thirteenth-century copy (see fig. 4.1 in chap. 4). Accordingly, this map will be discussed in connection with the cartography of the Middle Ages, as will the Madaba mosaic, an early Christian map of the sixth century.

Even though there are few surviving examples and the documentation is uneven, we can see that cartography in antiquity was well developed and diverse in terms of utilization, subject, scale, and materials used. We can now turn to the East, where important developments affecting cartography were taking place.

# THREE

## Early Maps of East and South Asia

The record of early cartography in much of the East is as fragmentary as that in the West, and for most areas even more so. There was contact between India and Mesopotamia at an early date, and influences, including mathematical and cosmological ideas, traveled in both directions. In 326 B.C. Alexander the Great crossed the Indus and found great civilizations at Taxila and elsewhere.[1] Earlier, following the rise of Buddhism in India in the sixth century B.C., there were religious, scientific, and other cultural exchanges between that country and China. Later there were exchanges in the opposite direction, especially after Buddhist pilgrims from China sought texts (including, perhaps, maps) in the Ganges Valley from at least the fourth century A.D. We will return to discuss the cartography of south Asia subsequently.

But it is appropriate here to review briefly the early geographical and cartographical contributions of the Chinese, both because they are important in their own right and because they exerted an influence on other parts of the Orient.[2] In the literature of China, we have evidence of geographic and cartographic activity of a much earlier date than that of the oldest surviving maps of this civilization. The earliest survey of China (Yü Kung) is approximately contemporaneous with the earliest reported mapmaking activity of the Greeks, that of Anaximander (sixth century B.C.). In the centuries following, there are remarkable parallels between the geographical literature of China and that of Greece and the Latin West (especially later Roman writers), indicating more than casual contacts between these cultures. In China, as in the West, we can identify anthropogeography; descriptions of the home area and of foreign countries; coastal and hydrographic books; urban as well as local topographic studies; and geographical encyclopedias.

It is assumed that maps, charts, and plans accompanied even very early examples of these geographical works. A specific reference in Chinese literature alludes to a map painted on silk in the third century B.C.;

27

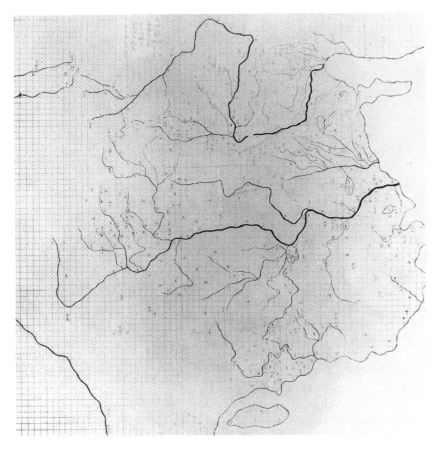

*Figure 3.1.* Early map of China, with rectangular grid, carved on stone.

the weft and woof of the material may, in fact, have suggested a map grid. We also learn that various rulers, generals, and scholars during the Han Dynasty (207 B.C.–A.D. 220) had a high regard for maps and used them for military and administrative purposes. Recently two maps were found during excavations of Han Dynasty tombs (ca. 168 B.C.) that are the oldest surviving maps of parts of China. They are drawn on silk and were preserved in a lacquer box. One is a regional map of a large area of Honan province with relief features, rivers, and settlements delineated and named; the other is a map of a garrison and its surrounding area on which even the loyalties of the neighboring villages are indicated.[3]

Apparently the rectangular grid (a coordinate system of equal squares), which is basic to much scientific cartography in China, was

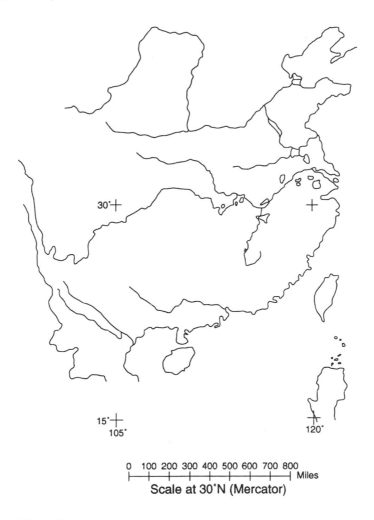

30°+

15°+
105°

+

120°

0  100 200 300 400 500 600 700 800
├──┼──┼──┼──┼──┼──┼──┼──┤ Miles
Scale at 30°N (Mercator)

*Figure 3.2.* Outline map of China from a modern chart, for comparison with fig. 3.1.

formally introduced by the astronomer Chang Heng, a contemporary of Ptolemy. The grid subdivides a plane or flat surface; although this figure was assumed for purposes of mapmaking, it must not be supposed that all scholars in China believed this was the shape of the earth. Indeed, we know that the Chinese used the gnomon and were aware of the continual variation in the length of its shadow in the long north–south extent of their own country—knowledge that presumably suggested to them a curving surface, if not a terrestrial globe. Their catalogs of eclipses and comets (including Halley's) are amongst the oldest and for some periods the only records we have of these phenomena.⁴ Early Chinese contributions to the cartography of the heavens include a star map (A.D. 310); the first celestial globe (A.D. 440); the earliest surviving manuscript star map (A.D. 940); and a planisphere (A.D. 1193). It is even claimed that they employed a "Mercator" projection for astronomical charts, but it was more likely a simple cylindrical projection.

In the third century A.D., we learn of a prime minister's younger sister embroidering a map to make the record more permanent.⁵ In the same century, Phei Hsiu, a minister of works during the Chin Dynasty (A.D. 265–420), outlined the principles of official mapmaking, which included the before-mentioned rectangular grid for scale and locational reference; orientation; triangulation; and altitude measurement. Unfortunately, none of the maps of Phei Hsiu has survived, but as in the case of lost Greek maps, modern scholars have attempted reconstructions of these cartographic works from written descriptions.

As the imperial territories of China increased through the centuries, maps of various scales were made of the enlarged realm. These works set the stage for further cartographic accomplishments, which may be exemplified by a map of China dated A.D. 1137 or even earlier. The map in question (fig. 3.1) has a regular rectangular grid with a scale of one hundred li (about thirty-six miles) to each square. It locates settlements and delineates the coastline and the major rivers of China—the Yangtze and Hwang Ho and their tributaries—in clearly recognizable form. (An outline map of this large area with details taken from a modern chart is provided for comparison in figure 3.2). The map, which is about three feet square, was carved in stone by an unknown cartographer of the Sung Dynasty; its purpose was to illustrate a much earlier geography based on the before-mentioned survey, Yü Kung. Needham and Ling justly assert that this map is "the most remarkable cartographic work of its age in any culture."⁶ Its portrayal of the coastline and drainage of China bears a remarkable resemblance to current delineations of the area and in this sense is better than any map, European or Oriental, until the period of modern systematic surveys. As with the publicly displayed Roman world map of Marcus Vipsanius Agrippa discussed in chapter 2 and many others in the modern world, this map of China was instructional in purpose.

Another Chinese cartographic milestone of about the same period is the earliest known printed map. It is assumed to have been made around A.D. 1155 so it predates the first printed European map by over three centuries. It is likely that earlier examples have not survived, since printing was invented in China in the eighth century A.D. and was used for scientific treatises in the following century. This early map (fig. 3.3),

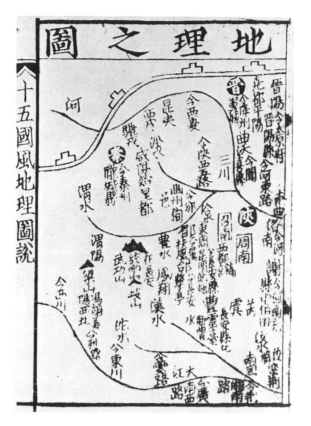

*Figure 3.3.* The earliest printed map, surviving from China, depicting a
portion of the western part of the country and showing part of the Great Wall,
rivers, mountains, and settlements.

which served as an illustration in an encyclopedia, is printed in black
ink on paper (which had been invented in China in the second century
A.D.), and it shows part of western China. In addition to settlements and
rivers, a portion of the Great Wall is indicated at the north. Both this
map and the one illustrated earlier (fig. 3.1) have north orientation—
that is, north is at the top of the map—which of course is now conventional in the West. The Chinese sometimes used orientations other than
this, as did different peoples with whom they had contact. (For example,
the Arabs, who settled on the coast of China before A.D. 750, characteristically made south-oriented maps.) A terrestrial globe is listed as part of
the inventory of a Chinese observatory in the thirteenth century, but
this was probably because of the influence of the Persian Jamal al-Din
ibn Muhammad al-Najjari (Cha-Ma-Lu-Ting).

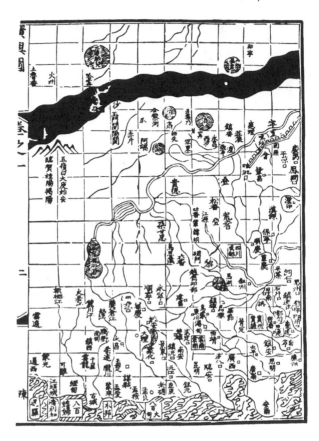

The culmination of indigenous Chinese cartography is found in the contributions of Chu Ssu-Pen (A.D. 1273–1337) and his successors, who established a mapping tradition that lasted until the nineteenth century. Chu, who built upon a scientific cartographic heritage extending back to Chang Heng and Phei Hsiu, like his predecessors made a manuscript of China with a rectangular grid. The reliability of the information on which his map was based was of the greatest concern to Chu, whose attitude is quite modern in this respect. The map was constantly revised and eventually enlarged, dissected, and printed in atlas form some two centuries after Chu's death (fig. 3.4). The treatment of the ocean with angry lines, common in Oriental cartography, perhaps suggest percep-

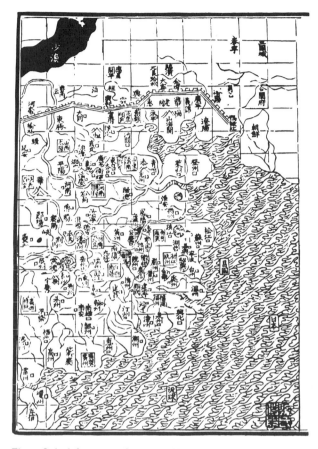

*Figure 3.4.* Atlas map of eastern China, showing part of the
Gobi (solid black area), the Great Wall, and the China Sea.

tion of the seas as a hostile environment. However, not all Chinese regarded the oceans in this way, as evidenced by the early Ming Dynasty
voyager Cheng Ho. With great fleets of junks he visited many countries
from Vietnam to East Africa between 1405 and 1435 using nautical
charts, a copy of one of which is preserved. The Chinese might have
made even more spectacular maritime discoveries had not these expeditions been officially discouraged after Cheng Ho's death.

Although this book is more concerned with maps than with the
methods used to produce them, it should be mentioned that, at least by
the time of Chu Ssu-Pen, Chinese cartographers knew the principles of
geometry and possessed instruments that would greatly facilitate their

mapping activities. The instruments included the gnomon, as mentioned previously, and a device similar to the *groma* of the Romans, with plumb lines attached. The Chinese also used sighting tubes and something akin to the European cross-staff for estimating height, as well as poles for leveling and chains and ropes for ground measurement. The odometer, or carriage-measuring instrument, by which distance is ascertained by the revolutions of wheels, is referred to in China at least as early as in Europe. Compass bearings, implying the use of the magnetic needle, seem to have been made by the eleventh century A.D.; it is assumed that the magnetic needle was transmitted westward to Europe shortly after this period.[7] Actually, a reference to a south-pointing chariot and a mounted magnetized needle goes back to the Wei Dynasty (third century A.D.), but we are not sure when this instrument was first used for mapmaking.

The Chinese also made maps of large areas beyond their own borders, but because these regions were of less importance to them and because of their (understandable) Sinocentric point of view, foreign countries tend to be minimized as a function of their distance from the culture hearth of the empire.[8] We have mentioned that Chinese cartography influenced that of other areas in eastern Asia. This was especially true for Manchuria and Korea, although the cartography of these areas is not without innovation. In the latter country, several documents point to the existence of maps prior to the Lee Dynasty, which had its origins in the late fourteenth century. Manuscript copies of "world" maps from this period indicate that Oriental knowledge of the West was greater at this time than the reverse, which is attributed to Arab, Persian, and Turkish contacts. As far as their own territory was concerned, the Koreans made increasingly detailed maps showing the direction of mountain ranges, the distance of towns from the capital, and the boundaries of political units, mainly for administrative purposes.[9] It appears that the transmission of printing from China came rather late, but a printed map from Korea antedates any European example.

Although China influenced Korean cartography directly, it was through Gyogi-Bosatsu (ca. A.D. 688–749), a Buddhist priest of Korean origin, that Japanese mapmaking was more immediately advanced.[10] He was a builder of roads, bridges, and canals and advised the rulers of Japan on mapmaking. We know of so-called Gyogi-type maps only through later manuscript and printed copies, which display the whole country divided into sixty-eight provinces with main roads from those provinces to the ancient capital of Kyoto indicated. In contrast to these small-scale maps are urban representations that exhibit great detail, such as that of Kyoto from A.D. 1199 (fig. 3.5). The gridded layout of the city is drawn in

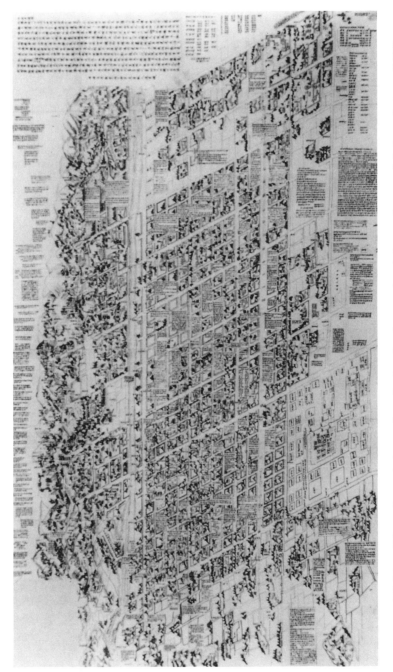

*Figure 3.5.* Map of Kyoto, Japan, in isometric projection showing temples, shrines, palaces, and drainage features (1199).

isometric projection (prefiguring some modern city plans) with temples, shrines, and palaces clearly marked. This tradition of urban mapping continued in Japan until comparatively recent times.

We began our discussion of early maps in East and South Asia with a reference to the Indus Valley civilization and noted the unfortunate lack of evidence of cartography in that area. There is a growing interest in the precolonial cartography of southern Asia: Pakistan, India, Sri Lanka, and the region from the Himalayan states to peninsular and insular Southeast Asia. However, in spite of the commendable work of Reginald H. Phillimore, Susan Gole, R. T. Fell, Joseph E. Schwartzberg, and others, the early cartography of this ancient, populated area is still in an embryonic stage.[11] No studies such as those of Needham and Ling for China or of Youssouf (Yūsūf) Kamal for Egypt and Africa exist for this large region. A major problem for Western scholars has been the difficulty in understanding the complex cultures of these Asian peninsulas and islands, and more pressing problems are being addressed by indigenous savants. Archives exist that scholars have not even visited—much less worked in—but that may contain original materials. A body of scholarship as large and detailed as that for Western mapping, involving many disciplines, is needed to uncover the riches of the cartography of this area.

In the absence of such primary studies, we can only speculate on the nature of the early cartography of South Asia from tenuous evidence. As in other areas, some drawings in caves—of which many examples exist—have been interpreted as maps. It also appears that cosmological diagrams of transcendent value to the inhabitants and analogous to those of other cultures (especially the Babylonian and Mayan) have been made in "greater" India since early times, especially by the Jains. Similarly, plans of temples, altars, and other structures were created by the deeply religious peoples of the subcontinent and the areas they influenced. A few of these exist on nondurable materials, such as palm fronds, suggesting that a larger number may not have survived. Most of this cartography, as well as city plans, is known to us through very late examples or reconstructions.

A copy of a map from the late Moghul (Mughul) Empire is reproduced as figure 3.6. It shows a large area of northwestern India, Pakistan, and Afghanistan; the top of the map is to the northeast. A square portion near the center is blank, having been destroyed by insects or climate. Extending from this missing area is the Indus, with its distributaries at the left margin of the map. The great cities of northern India are shown and named within pictorial rectangles (Delhi-Shahjahanabad, Agra-Akbarabad, and so on) on the bottom margin. The hills around Ajmer

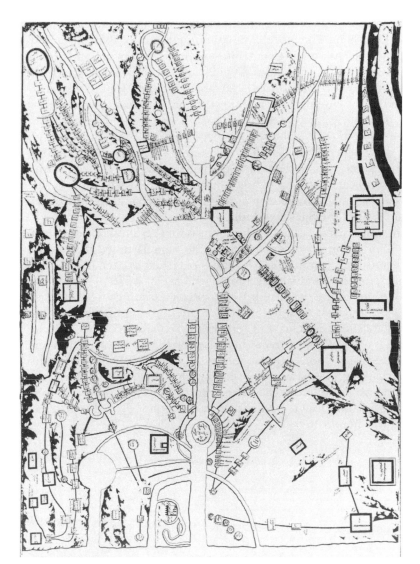

*Figure 3.6.* Map of part of the Moghul Empire that is now Pakistan, Afghanistan, and northern India.

are drawn in profile as viewed from different prospects in the city. Routes are marked and distances between important places noted. Three of the principal tributaries of the Indus in the Punjab, the land of the five rivers, are shown—the Beas, Ravi, and Jhelum—but the Sutlej and Chenab are missing, probably because of the blank area. The Ganges is at the bottom of the map. The more mountainous country of Baluchistan, Afghanistan, and Kashmir is toward the top, indicated by uplands (dark profiles) surrounding such cities as Quetta, Kabul, and Srinager in Kashmir (Kushmeere). It was to the latter area that the Moghul emperors journeyed in the summertime to enjoy the cooler climate and the great gardens that they created there. Maps such as these would have been of immense value to the Moghuls in their travels for pleasure and in their military campaigns. They were also of interest to the British when they began their monumental topographic surveys of India in the eighteenth century, as indicated by the English transcriptions of place-names beside the Persian originals on the map as well as by the retention of this map in the archives of the Survey of India.

Some remarkable "cartographical curiosities" from Asia are extant, such as Chinese incense burners from the Han Dynasty, the earliest three-dimensional terrain models, and—of a much later date—Japanese plates, mirrors, and fans decorated with maps. We shall now return to the West, including the Islamic world, which was enormously influential in South and East Asia, but not before mentioning that when Jesuit fathers established residence in China in the sixteenth century, the cartographic record of this area was made available to the Europeans and incorporated into their regional and world maps. Although after this time mapmaking in Asia was overwhelmingly influenced by European techniques, we do well to recall the extraordinary cartographic accomplishments in the Orient, particularly as exemplified by the works of early Chinese mapmakers.

# FOUR

## Cartography in Europe and Islam in the Middle Ages

For the sake of convenience, the term *Middle Ages* is used in this chapter, but obviously this has more relevance to Europe than to Islam. The earlier part of the medieval period—the so-called Dark Ages, from ca. A.D. 450 to 1000—is not now considered to be so dark as was formerly thought to be the case, but little Western cartography survives from this half-millennium. One extant map, however, is the Madaba mosaic, ca. A.D. 590, which takes its name from the Jordanian town where it was found in 1889.[1] Sometimes described as the first Christian map, it served as part of the floor of a church before being restored. Though it is incomplete, remaining fragments suggest that it focused on western Asia but included, at its extremities, the Nile Delta and the Black Sea. It was made by Byzantine mosaicists and shows the city of Jerusalem in remarkable detail.

Another map that has its origins in this period, or even earlier, is the *Tabula Peutingeriana* (Peutinger map), alluded to in respect to Roman cartography in chapter 2, which takes its name from a sixteenth-century owner of the map, the humanist Konrad Peutinger of Augsburg.[2] This map is a large manuscript on parchment, roughly one foot high with an overall length of more than twenty feet. It was originally a roll but has been dissected into twelve sections, the first of which, now lost, is thought to have shown most of Britain (a small part of which still exists on section 2), the Iberian peninsula, and adjacent parts of North Africa. The surviving eleven sections illustrate, in a highly diagrammatic fashion, the area from extreme eastern England through the Mediterranean to India. It is presumed that the information is from Roman itinerary maps of the first century A.D. As it has come down to us, the map appears to be mainly fourth-century work with some additions made as late as the sixteenth century, when it was first published.

Figure 4.1 shows a small portion of the Peutinger map, where sections 6 and 7 adjoin, focusing on Sicily and the "boot" of Italy. As can be

*Figure 4.1.* Simplified rendering of a small section of the Peutinger map, focusing on the boot of Italy and Sicily, redrawn by Noel Diaz. Because of the greatly reduced scale of the map, lettering (which forms an important feature of the original) is omitted.

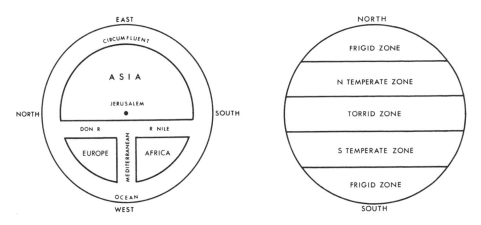

*Figures 4.2 and 4.3.* Diagrams of the T-O concept (*left*)
and the zonal concept (*right*) of the earth.

seen from this sample, areas on the map are very elongated in the east–
west direction. It is not developed on a systematic projection, but approximate distances between settlements are written on the map. The
roads are delineated predominantly by straight lines, often with curious
jogs; routes are in red, while the sea is indicated in greenish blue. The
map is enriched by the rendering of mountain chains (in profile), buildings, and, in some cases, persona at important centers such as Rome,
where a figure holding an orb, a shield, and a spear sits enthroned. The
major roads of the empire emanate from the "Eternal City."

For a compilation such as the Peutinger map, as with early Chinese
maps, a spherical earth need not be assumed. This is especially true of
maps of small areas, such as city plans of Rome and Constantinople,
which were part of the library of Charlemagne (A.D. 742–814). These
maps, and also a world map belonging to Charlemagne, were engraved
on silver tablets, but we know of them only through references in literature. There is no proof that, in general, medieval people believed in a
flat earth. In fact, we know specifically that a number of influential savants of this period accepted a globular world. Nevertheless, there were
those who argued against this concept, as well as the related idea of inhabitants in the Antipodes.[3] In the Middle Ages, the world was represented on maps by various shapes: irregular, ovoid, rectangular, cloak-
shaped, and circular. The most common shape was the circular, disc, or
wheel form, a descendant of Greco-Roman cartography, of which two
distinct types can be recognized: the T-O and the climatic zonal forms
(figs. 4.2 and 4.3) They relate to the geographical ideas of Macrobius
(ca. A.D. 400), Orosius (early fifth century A.D.), and Isidore of Seville

(ca. A.D. 600), who provide links between antiquity and the later Middle Ages. A simple circular map by Isidore later became the first printed European map in 1472 (see fig. 5.1 in chap. 5).

The T-O *(orbis terrarum)* type usually has east, or the Orient, at the top (hence the term *orientation*) with Asia occupying the upper half. Asia is separated from Africa by the Nile and from Europe by the Don (Tanais). Together, these two rivers form the top of the T, while the Mediterranean, which separates Europe from Africa, forms the upright segment. The whole is surrounded by a circumfluent ocean—the O. This cartographic form embodies geographical concepts from antiquity that we have encountered earlier. These include the notions of three continents—a widely accepted idea—and of land covering most of the earth with separating seas and a circumfluent ocean, exemplified by the ideas of Crates. T-O maps also satisfy Christian theology by giving Jerusalem a central position on the earth.[4] The names and sometimes the portraits of the sons of Noah, the assumed ancestors of the peoples of the three then-known continents—Japhet (Europe), Shem (Asia), and Ham (Africa, particularly Libya)—appear in their correct locations on many of these maps.

In the zonal type of map, the debt to Greek science is more explicit, and the examples frequently have orientations other than to the east. As the Ancients were concerned with the extent of the inhabited earth or ecumene *(oikoumene)*, which they attempted to delimit with latitude *(climata)*, so medieval scholars were interested in the area of human occupancy, but for theological reasons. Both the T-O and the zonal types of circular world maps, which often accompany medieval manuscripts, are also found in the cartography of Islam; in some instances, both concepts (and others) are combined on a single map in Christian and in Muslim examples.

The ultimate expression of the circular world map of the Middle Ages is found in the Ebstorf and the Hereford maps. Both of these maps are examples from the late thirteenth century, and they are similar in conception, though quite different in detail. These *mappaemundi* seem to have served as altarpieces (or perhaps were hung behind the altar), in the German monastery church of Ebstorf in the one case and at Hereford Cathedral, England, in the other. Unfortunately, the Ebstorf map was destroyed during World War II, but good color copies of this large manuscript map, which was about eleven and half feet in diameter, survive.[5] The Hereford map (fig. 4.4), which is drawn on vellum—possibly a bullock's skin—is also large, measuring five feet three inches high and four feet six inches wide. It remains one of the treasures of the Hereford Cathedral since it was miraculously saved recently from the auctioneer's

*Figure 4.4.* Photograph of the original manuscript Hereford world map.

gavel. It is a type of map probably found in a number of the great religious houses of Europe during the Middle Ages.

The Hereford world map was formerly thought to be something of an oddity and of little geographical value, but as a result of later research this opinion is being revised.[6] According to an inscription on the map, which requests the prayers of the viewers for the author, it was made by Richard of Haldingham and Lafford (Richard de Bello). Richard may have brought the map with him when he came from Lincoln to Hereford toward the end of the thirteenth century, though some believe that

*Figure 4.5.* Detail of the upper part of the Hereford world map.

it was made in Hereford by Gervais of Tilbury, the author of the Ebstorf map. Both represent a modified form of the T-O map and have an affinity with earlier examples, including those of Macrobius, Orosius, Isidore, and Henry of Mainz (A.D. 1100). The Hereford map is based on classical itineraries and later sources and can be regarded as a summary of the geographical lore, secular and sacred, of the Middle Ages. It depicts mythical creatures and abnormal people from the fabulists, but it also contains new information derived from medieval commercial journeys, pilgrimages, or Crusades, especially in the European section. In its coloring and calligraphy, as well as its view of the earth, the Hereford *mappamundi* beautifully expresses the feeling of the later Middle Ages, which reached its highest architectural development in the Gothic cathedral. In fact, there are close relationships between the symbolization used in these two forms. For example, the representation of Christ in judgment above the circumfluent ocean on the Hereford world map (fig. 4.5) is reminiscent of the tympana in contemporary ecclesiastical architecture, which originally were often colored. The perfect celestial world is set apart from and above the imperfect terrestrial world. Although some savants in the Middle Ages, such as the Venerable Bede and Roger Bacon, investigated physical phenomena, there were at times proscriptions against geographical explorations, including the measuring of the depths of the sea.

It has been debated whether such works as the Hereford *mappamundi* were intended to aid travelers or were inspirational pictures like the stained glass windows of the cathedrals. No doubt they served both functions, and some pilgrims who looked at them probably suggested changes based on their own travels.[7] However, there were also medieval maps more specifically designed for the assistance of travelers. Thus, at the scriptorium of St. Albans, the thirteenth-century monk Matthew Paris made not only a *mappamundi* and a map of Great Britain but also a strip map showing pilgrimage routes within England and another showing routes from London to southern Italy. Such strip maps resemble the Peutinger map and anticipate modern road strip maps in that routes are indicated by straight lines from place to place, with no special attention paid to orientation. The representation of Britain by Paris can also be thought of as an itinerary map, although the routes are not specifically delineated. Said to be the oldest map of a European country, Paris's map of Britain (fig. 4.6) shows a compressed Wales and Scotland (beyond the Roman Wall), rivers, and fortified cities (named). A similar but more spatially recognizable map of Britain from a century later, and on which major routes are depicted, is the Gough map, so named after a later owner, the eighteenth-century antiquary Richard Gough. The Gough map might have been made for administrative purposes.

While these developments were taking place in the religious houses and courts of medieval Europe, mapmaking was progressing elsewhere. We have already discussed contemporaneous mapping activities in the Far East, but scholars from the Middle East also contributed notably to cartography in this period.[8] After the fall of Babylon, science in the arid lands of southwestern Asia appears to have been strongly influenced by India. This is particularly true before the reformation of the Arabic alphabet and the translation of Ptolemy's works, the *Almagest* (astronomy)

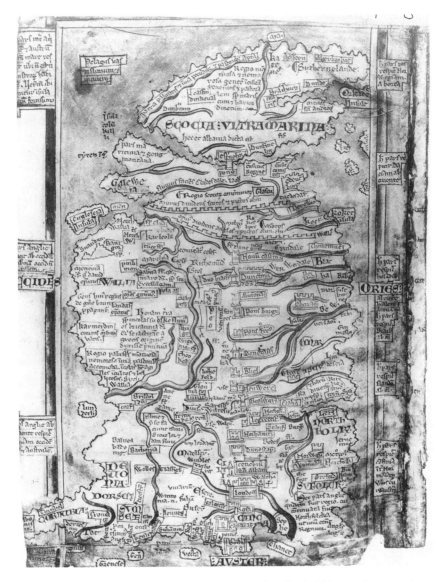

*Figure 4.6.* Manuscript map of Britain by Matthew Paris (thirteenth century).

and *Geographia* (cartography), into Arabic in the ninth century A.D. In the next century, Ptolemy's maps, possibly constructed from instructions in the *Geographia* rather than from originals, became available to the Arabs. Following this, two distinct developments ultimately affecting cartography unfolded in Islam: (1) the determination of the latitude and longitude of places on the earth as part of an increasing emphasis on astronomy; and (2) geographical descriptions arising from the extensive land and sea journeys undertaken for conquest, administration, or trade. The first of these developments, however, did not always have an immediate effect on the second.

Although at first Arab scholars seem to have accepted Ptolemy's astronomical works, in time they criticized and improved upon them. The length of the Mediterranean Sea, given by Ptolemy as 62 degrees, was reduced to 52 degrees by Al Khwārizimī (ninth century A.D.) and further reduced to its correct figure of 42 degrees through the observations of Al-Zarqālī (twelfth century A.D.). We have noted the impact of Islamic, including Persian, science on China and India (whence came the idea of zero). The debt was not unidirectional, as the Middle East was the beneficiary of, among other things, both papermaking and printing from China. In respect to the latter invention, hand lettering (calligraphy) seems to have been preferred to printing in both areas for a long time. Following the lead of China, and of Greece and Rome, civilizations to which it was especially indebted, Islam can claim production of celestial charts from the first century after the establishment of this faith (seventh century A.D.). Planisphere or celestial charts based on systematic or formal projections are a strong tradition in Islam, as is a related instrument, the astrolabe or star measurer, some of which were designed to find the direction of Mecca. The astrolabe, usually made of metal, is a stereographic projection of the celestial sphere showing great circles accurately; manipulating its moving parts allows the elevations of the sun and stars to be measured.[9] Although invented earlier by the Greeks, astrolabes of Arab provenance from the ninth to the nineteenth centuries are extant. Similarly, cosmological diagrams (always with a geocentric focus), zodiacal signs, and spheres of the four elements and of *climata* are part of the Islamic tradition inherited from, but modified after, Greek and Roman models. The oldest surviving globe, a celestial sphere, is also Islamic, made by the Persian Muhammed Ibn Mu'aiyad al-'Urdhi and dated from internal evidence at 1279.

It has been said, perhaps unfairly, that the Arabs were better astronomers and geographers than cartographers. There are a number of geographical map types from early Islam, as evidenced by surviving examples of the rich cartography of this culture: religious topographies, analogous to those of other areas, such as medieval Europe; world maps, typically disc-shaped, often stylized and with a circumfluent ocean; regional maps, as of part of the Nile Valley and smaller areas; military maps, including siege plans; urban views, either in plan or bird's-eye aspect; and itineraries.

Islamic cartographers whose names and compilations are well known to us include Al-Iṣṭākhri and Ibn Haqal (both of the tenth century Balkhi, Irani School), Ibn al-Wardī, Al Khwarizimi, and Al-Idrīsī. Many of their maps are south-oriented and the symbols highly conventionalized; there was often great reliance by later cartographers on earlier ones. Some of these cartographers were travelers, some were theoreticians, and some combined both roles. The latter was the case with Abū 'Abd-Allāh Muhammad al-Sharif al-Idrīsī (ca. 1100–1165 or 1166), who, after undertaking extensive travels from his native Morocco to Asia Minor, was invited to Sicily by its enlightened Norman king, Roger II. In Sicily, Idrīsī engaged in geographical writing and in the compilation of maps. He made a circular world map (fig. 4.7) with curved parallels that is superior to contemporaneous European maps of the same genre in a number of respects. His most important work, however, is a large rectangular world map in seventy sheets, known as the *Tabula Rogeriana*. Figure 4.8 is a reproduction of a page of Idrīsī's atlas showing the Aegean Islands, while figure 4.9 is a redrawing of a large section of Idrīsī's world chart of 1154, the earlier and generally better of two such maps that have come down to us through the work of copyists. It is instructive to compare Idrīsī's south-oriented map with the Hereford map (fig. 4.4) of approximately the same time. Obviously, the *Tabula Rogeriana* is less stylized, and it incorporates new information supplied from the travels of Idrīsī himself and many others. Some of the cartographic work executed at the Norman court in Sicily, including a map engraved on a silver tablet, has perished, but the surviving record demonstrates the originality of Idrīsī's contributions, which continued to be important in the Arab world for centuries after his death. The extent of Ptolemy's indirect influence on Idrīsī (who also had access to Balkhi school maps) has been speculated upon but takes nothing away from latter's accomplishments.

Some Muslims were great travelers by land and sea, the Sindbad tradition having arisen from their epic navigational feats in dhows of sewn-plank construction fitted with triangular (lateen) sails. In such vessels and also by land, they extended from their North African and Middle Eastern homelands to India, Southeast Asia, and China to trade, proselytize, and settle; the Chinese, by contrast, were content to return to their own country following their voyages in the same period. Not surprisingly, Muslims developed sophisticated navigational techniques that later proved to be of value to Europeans, as will be indicated below. The greatest of the later Islamic travelers was Abū 'Abd-Allāh Muahammad Ibn-Baṭṭūṭa, a younger contemporary of Marco Polo (1254–1324), the most celebrated European traveler of this time. Both journeyed over

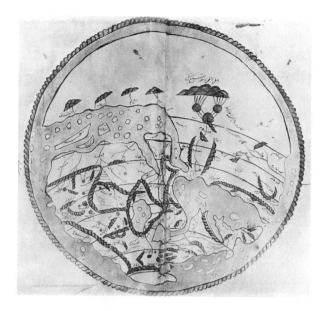

*Figure 4.7.* Arabic zonal world map, with south orientation, by Idrīsī.

*Figure 4.8.* Peloponnisos, Kikladhes, and Kriti (Crete): a small section
of Idrīsī's world map, with south orientation, in atlas form.

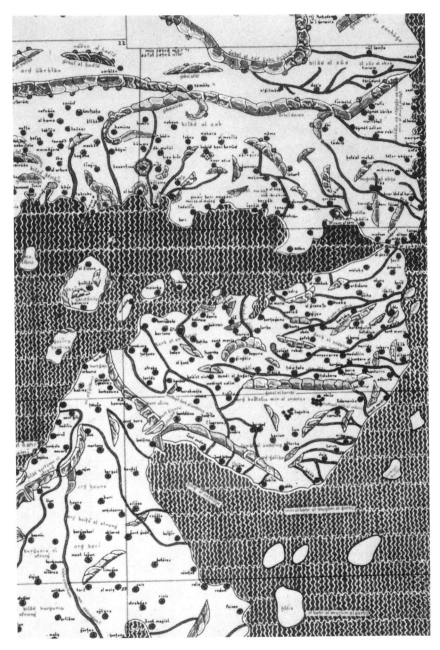

*Figure 4.9.* Detail of Idrīsī's world map, with south orientation.

some of the same areas from Europe to China, both supplied data for maps, and both were originally from the Mediterranean, where Muslim, Jewish, and Christian learning, including cartography, mingled.

We mentioned the magnetized needle in the discussion of Chinese cartography and speculated on its use in mapping in that area and on the transmission of this useful instrument to Europe either by sea via the Arabs or by way of the Silk Road across central Asia. In the later Middle Ages, there was great interest in Europe in the properties of the lodestone, and experiments involving magnetism were performed by Roger Bacon (1219–92), who also devised a globular map projection. Various attempts were made to satisfactorily mount the needle, and apparently this was accomplished in Amalfi, Italy, at the end of the thirteenth century. In the Mediterranean, the compass—consisting of a box containing a pivoted, magnetic needle mounted over a card on which sixteen, and later thirty-two, directions were painted—came into use among sailors. Because the directions were named for winds, following the practice of antiquity, such geometrical constructions are known as *wind (compass) roses*.[10] With the help of the magnetic compass, great progress in mapping and in navigation was possible, and a new cartographic form presumably related to this development appeared in the later thirteenth century: the portolan chart. We know that before this date the magnetic needle was in general use in the Mediterranean.

The origin of the portolan chart is obscure, but it seems to be an obvious extension of the descriptions found in pilot books *(portolani)*. That the earliest surviving examples are well developed suggested that even earlier portolan charts may be lost. They are typically drawn on a single sheepskin and oriented according to magnetic north, although other orientations are possible. Since their purpose was to aid navigators, shorelines are emphasized and in early examples little geographical information appears on the land. Characteristically, portolan charts show the Mediterranean and Black Sea coasts with remarkable accuracy, though with curiously stylized symbols, emphasizing headlands. Place-names are lettered perpendicular to the shore, the more important in red and others in black. Striking features of these maps include the representations of compass or wind roses with rhumb lines emanating from them, crisscrossing the charts. To use such a chart, a pilot would first lay out a course from a port of departure to a port of arrival with a ruler; then the line most nearly parallel to the ruler would be traced back to the "parent" compass rose to identify the required bearing (rhumb) on which to sail. We know of the method for constructing portolan charts from a diagram in the 1318 *Atlas* of Pietro Vesconti (who worked in both

Figure 4.10. *Carte Pisane,* an example of a portolan chart.

*Figure 4.11.* The shore of southern Italy and Sicily (represented
by the solid line), redrawn from the *Carte Pisane;* the dotted line is the
same shore as it appears on a modern map.

Venice and Genoa), which showed a circle with sixteen equally spaced
compass roses. There is an almost conventional color scheme for the
lines extending from these: black for the eight principal winds; green
for the eight half winds; and red for the sixteen quarter winds.[11]

To illustrate this cartographic form, we have selected one of the
many surviving portolan charts, the famous *Carte Pisane,* or Pisan chart
(figs. 4.10 and 4.11). This oldest portolan chart (ca. 1290) is, as it name
suggests, of Italian (Pisan or Genoese) origin; like most others from Italy,
it shows only the Mediterranean and Black Sea region. It is the first map
to have a graphical scale (to the east in a circle on the "neck" of the
parchment) subdivided into fifties, tens, and fives. The oldest *dated* portolan chart is that of Giovanni da Carignano, made in Genoa in 1310.
However, the maritime cities of Italy were not the only centers of chartmaking at this time. Portolan charts embracing the Mediterranean,
northern Europe, and others parts of the world were made by Catalan
(including Jewish) cartographers of Majorca and Barcelona in the service of the kings of Aragon.[12] Some scholars affirm the Catalan origin of
the portolan chart, but, at all events, in that area coastlines were extended and information from Arab and other sources was added so that

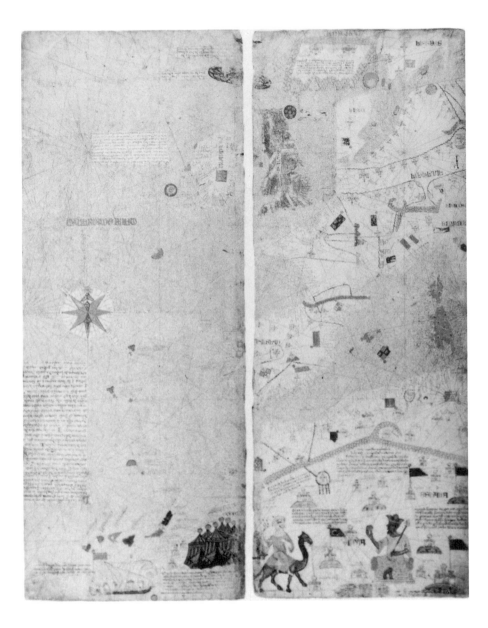

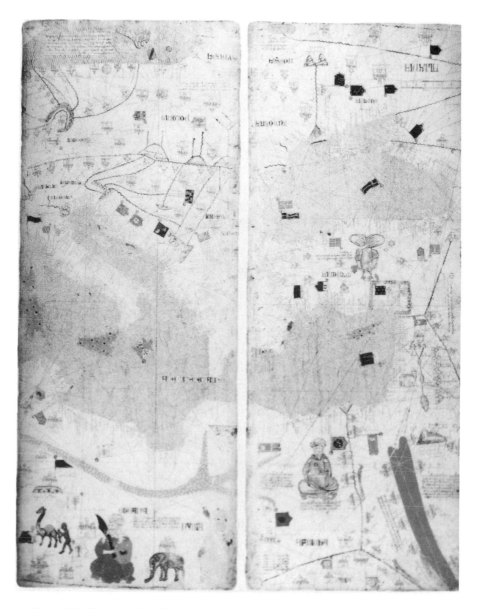

*Figure 4.12.* Four panels of the portolan-style Catalan map by Abraham Cresques (1375) showing Atlantic islands, Ireland and Britain, Western Europe, and North Africa (with Prester John, seated), to the Red Sea and Central Asia.

portolan-type world maps eventually developed. The famous and very rich Catalan map (1375) of Abraham Cresques is one example of this process (fig. 4.12). Among the iconography on this map is the figure of Prester John, the Christian priest-king beyond the pale of Islam whom the Europeans believed would help them overcome their enemies if contact could be made. Over the centuries Prester John's realm was said to be located first in central Asia and later in Ethiopia.

The gradual reconquest *(reconquista)* of Spain and nearby offshore islands during seven hundred years of Muslim occupation in Iberia left many Jewish scholars and artisans in Christian communities. Some of these were mathematicians, astronomers, instrument makers, and chartmakers. Among the latter were Abraham Cresques (1325–87) and his son Jefuda, residents of Majorca. To the senior Cresques is credited the portolan-style *mappamundi* showing the Old World from the Atlantic to China, including information derived from the reports of travelers such as Marco Polo. The map was presented by King Pedro IV of Aragon to Charles VI of France in 1381 and is now one of the treasures of the Bibliothèque Nationale in Paris.

As we shall see later, there are other connections between the navigational expertise of Europe and of Islam, but specifically the portolan tradition was transmitted to the Islamic world by a medical doctor, Ibrāhīm Al Mursī, through his manuscript chart of 1461 (865 of the Islamic calendar). Although the map is very similar to Genoese and Venetian portolan charts of the period, Dr. Al Mursī made considerable additions in Islamic territories, especially North Africa and the Levant. Fortresses along the Danube that the Muslims were attacking at the time, including Esztergom, are prominently featured. Al Mursī was originally from Murcia in southern Spain, but it was in Tripoli, where he practiced medicine, that he made his chart. It is on a gazelle hide, on the neck portion of which he drew the Islamic calendar. He decorated the map with arabesques and signed and dated it.[13]

Several circular and other world maps made in the fifteenth century were influenced by portolan charts: the Leardo *mappamundi* (1448); the Genoese world map (1457); and the culmination of such cartographic compilations, the Fra Mauro map (1459). Because of the circumfluent ocean on some late medieval maps, a navigable route from Europe to the Indian Ocean by way of southern Africa—denied on early representations by Ptolemy—appears to be feasible. It is possible that such a voyage had been made in antiquity as reported by Herodotus (ca. 450 B.C.), but this presupposes an Africa much smaller than in reality and a voyage of several years, allowing time for planting and harvesting along the way.

In addition to world maps, there survive from the late Middle Ages bird's-eye views of various cities, such as Constantinople by Cristoforo Buondelmonte (ca. 1420) and Jerusalem by Bernard von Breydenbach (ca. 1480), a genre that was to be greatly expanded later and that continues to the present time. Plans of local areas, cities, and churches exist, including the plan of the monastery of Saint Gall (ca. 830), which is remarkable for its consistency in scale, orthographic point of view, and very early date.[14]

We can now move into territory perhaps more familiar to most people: the maps of Renaissance Europe.

# FIVE

## The Rediscovery of Ptolemy and Cartography in Renaissance Europe

The age of European cartographic ascendancy began in the later Middle Ages with the development of the portolan chart and the rise of Muslim, Jewish, and other influences in the Mediterranean and the Atlantic fringes of the continent. Continued advances in these areas during the Renaissance spread inland and north of the Alps to produce a great flowering of map expertise that has been the main focus of the history of cartography until quite recently.[1] It is true that there was considerable interest in maps in Europe in the medieval period, as shown in chapter 4, but this was overshadowed by events that took place in the period immediately following, including the transmission and translation of Ptolemy's *Geographia;* the invention of printing in Europe; and the voyages of Europeans overseas.

The cartographic work of Claudius Ptolemy was discussed in chapter 2 in connection with contributions from antiquity, and his influence on Islamic science was considered in chapter 4. When the Turks expanded westward and reached Byzantium (Constantinople; now Istanbul) in the fourteenth century, refugees fled from that city, taking with them various treasures, including the Greek texts of Ptolemy's *Geographia*. These manuscripts reached Italy and by 1410 were translated into Latin in Florence (Firenze), where geography and mathematics, as well as art, were important fields of study.[2] Thus manuscript copies of the Latin translation of the *Geographia*—first without maps, then with regional maps, and eventually with world maps—became available. The influence of Ptolemy's work on Western cartography would be difficult to exaggerate. It is only necessary to compare the best medieval *mappaemundi* with the Ptolemaic maps to appreciate the general superiority of the latter, at least in terms of the delineation of geographical features. This did not prevent European scholar-cartographers from modifying the representation of certain areas and features of Ptolemy's maps. For example, the map of Scandinavia by the Dane Claudius Clavus (1427) was used to update the world map of Ptolemy printed in Ulm in 1486 (discussed below). In fact, Ptolemaic maps were both the starting point and the model against which progress in geographical discovery came to be measured.

The importance of printing to cartography lay not only in the reduction of the cost of maps (actually, in some circumstances, prints might be more expensive than hand-drawn examples) but more especially in the ability to produce essentially identical copies, "the exactly repeatable pictorial [graphic] statement."[3] We have already considered the prior invention of printing in the Orient and its application to cartography as early as A.D. 1155, but apparently a different method of printing was invented independently in Europe and owed nothing to Oriental practices. The first European printed maps date from the last three decades of the fifteenth century. These incunabula include simple woodcut reproductions of various T-O and zonal world maps, engraved copies of Ptolemy's world, and sectional maps from both Italy and Germany.[4] Landmarks in the history of European map printing include the earliest map printed in Europe, a simple woodcut of St. Isidore's T-O map, printed in Augsburg in 1472 (fig. 5.1); and the earliest printed Ptolemaic atlas, twenty-six maps printed from engraved copper plates, published in Bologna in 1477. Actually, maps from woodcuts were on the whole more numerous than those from engraved metal plates in the first years

*Figure 5.1.* T-O world map from Isidore of Seville's *Etymologiarum,* the first European printed map (1472).

of European map printing, but gradually copper engraving overtook the woodcut and remained the prevailing method of map reproduction until the nineteenth century. Other landmarks include the first two-color printing, using red and black for different classes of geographical names in the Venice edition of Ptolemy of 1511; and a woodcut, three-color, printed map of Lotharingia in the 1513 edition of Ptolemy, published in Strasbourg and edited by Mathias Ringmann. The abandonment of woodblock printing in favor of copper-plate engraving, which does not lend itself so readily to color printing, led to the virtual demise of the colored, printed map for some three centuries.[5] During this period, hand coloring of black-and-white copper-plate prints became an important activity in European cartographic establishments, with women apparently playing an important role.

With this background, let us look at the Ulm edition of Ptolemy's world map of 1486, a woodcut print (figs. 5.2, and 5.6b). It is developed on a modified cloak-shaped *(chlamys)* projection—an ingenious solution to the problem of representing the all-side-curving figure of the globe (or, at least, a large part of the earth) on a flat surface. This is the most complicated of the three projections devised by Ptolemy, who also used a simple pseudoconic form for world and sectional maps. The map has a regular grid of numbered meridians and parallels, a new feature, and, on the western edge latitudes are expressed in hours of the longest day of the year *(climata)*. Beyond the margin (neat line) of the Ulm map, twelve wind blowers indicate direction. This is an elaboration of the concept of four wind directions of antiquity, a number that was increased to eight (as used on early magnetic compasses to divide the circle of the horizon), to sixteen, and eventually to thirty-two. Other features of the Ulm map include rivers, lakes, and mountain ranges.

Without going into great detail concerning land and water relationships on Ptolemy's world map, we should note that the world then known to the Europeans to be inhabited (Old World) extends about 180 degrees, halfway across the sphere in the northern midlatitudes, whereas in reality it covers only about three-eighths of this distance. Other features include an enclosed Indian Ocean with no apparent sea route around southern Africa to India, a comparatively good representation of the Malay Peninsula, a truncated India with an exaggerated Ceylon (Taprobane), and no east Asian coast. The delineation of the Mediterranean is only fair in comparison with existing portolan charts. This region and that of northern Europe, which is poorly represented in Ptolemy's maps, were soon to be improved. As the Europeans extended their sphere of influence overseas, the conception of the Ptolemaic world was shortly to be radically altered. Features similar to those noted above can

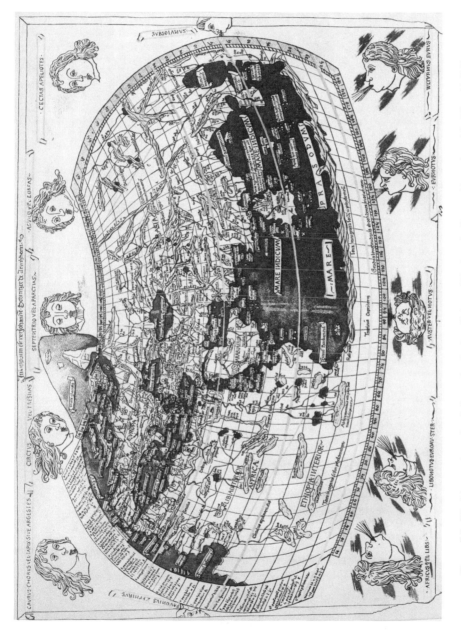

*Figure 5.2.* World map (woodcut) from the Latin edition of Ptolemy's *Geographia*, printed in Ulm (1486).

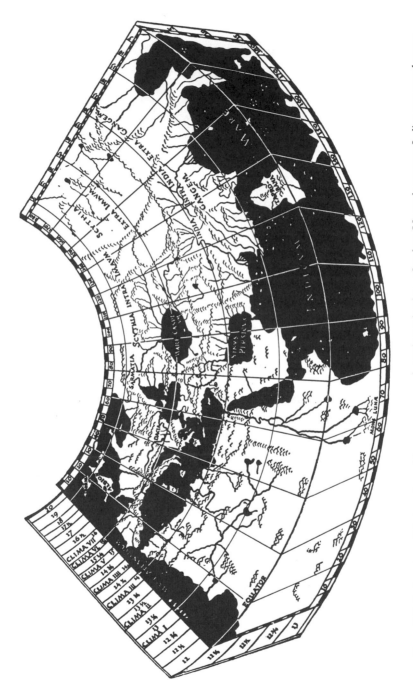

*Figure 5.3.* Redrawing of Ptolemy's world map on a pseudoconic projection, after late-fifteenth-century Italian engraved maps.

be seen on another world map plotted on Ptolemy's first and simpler projection. This one is based on copper engravings from late-fifteenth-century Italian examples of Ptolemy (figs. 5.3 and 5.6a), but a thirteenth-century Byzantine manuscript map of the inhabited world on this projection also exists. A version of this map was printed in the *Nuremberg Chronicle* by Hartmann Schedel (1493), the most elaborately illustrated book of the fifteenth century. This woodcut map has portraits of the three sons of Noah and, on the margins, bizarre figures with diseases and deformities from Pliny, Solinus, and others. Thus, anthropomorphically, this work is a throwback to the Middle Ages, but technically it heralds the age of the printed atlas.

The rediscovery of Ptolemy's works in the West coincided with the beginning of the age of the great European overseas "encounters"; the two events are, in fact, related. This exploring activity was formally initiated by Prince Henry of Portugal (b. 1394), later called the Navigator.[6] After 1419, when he settled in the Algarve, the most southwesterly province of Europe, Henry enlisted the aid of all who could assist him in his work—sailors, shipbuilders, instrument makers, and cartographers. These last included a "master of sea charts," presumably Jafuda Cresques, son of Abraham Cresques from Majorca, who brought the Catalonian-Jewish portolan tradition to Portugal. In addition, one of Henry's older brothers, Prince Pedro, visited Italy—including, in 1428, Florence—for the purpose of collecting maps and new geographical information.[7] Specific borrowings from Islam during this period include the Arabic number system and the lateen (triangular) sail, which allowed vessels to sail closer to the wind than the square sail used in Europe since classical times. Lateen sails were fitted to caravels of the Portuguese for their early explorations.

Meanwhile, under Henry's auspices, various expeditions rediscovered the Azores, which appeared on maps he had received, and probed southward along the west coast of Africa. The Cape Verde Islands were reached and Islam outflanked before Prince Henry's death in 1460. By the end of the fifteenth century, America had been discovered by Europeans, and the sea route to India was found. During the next three hundred years, most of the coasts of the world were visited and at least roughly mapped by the explorers of Portugal, Spain, Italy, Holland, France, and England (Britain). Through exploitation of overseas areas, Europe grew from a poor peninsula of Eurasia in the Middle Ages to the most influential area of the world in the seventeenth, eighteenth, and nineteenth centuries.[8] Exploring activities produced a great body of information that, sooner or later, was added to compiled maps. Much of the history of cartography has been concerned with the "unrolling" of the world map, and there is a vast literature dealing with European geographical exploration and its cartographic representation.[9] In fact, the two subjects are so closely interrelated that it can be said that a place is not really discovered until it has been mapped so that it can be reached again. Although there are other ways of recording geographical data, the map is the most efficient and visual method.

A cartographic work that has aroused much controversy in recent years is the Vinland map, claimed to have been drawn "in the 1440s." Supposedly for the first time, it charted Helluland, Markland, and Vinland (part of the east coast of North America), discovered by Bjarni Herjolfsson (ca. A.D. 985) and Leif Eriksson (ca. A.D. 1000), as an inscription on the map testifies.[10] Based on chemical analysis of the ink used, most now affirm the Vinland map to be an elaborate forgery and, in fact, a twentieth-century compilation of voyages described in early sagas, but there are still those who aver that it is genuine. Others are of the opinion that the Eskimos (Inuit) had much greater cartographic skill than the Norse of this period and that some of their data might be recorded on European maps.[11] However, our concern here with the geographical exploration of the Europeans is only to help us better appreciate the cartography of the Renaissance, the last quarter of the fifteenth century and the first three-quarters of the sixteenth century—one hundred years during which the European perception of world land and water relationships changed more than it did in any comparable period. The few maps that have survived by explorers from this period are, characteristically, sketches of small areas, and it is more instructive for our purposes to illustrate growing knowledge of the world with maps by well-known cartographers who incorporated the findings of the discoverers in their work.[12] Typically, these more general maps also better illustrate developing cartographic methodology than do those of the explorers themselves.

For example, the then-new discoveries of the Portuguese, notably those of Bartolomeu Dias at the southern cape of Africa, were shown on a Ptolemaic-type manuscript map made in Florence in 1489 by the German cartographer Henricus Martellus Germanus, which had to be extended to the south to accommodate the new data. Dias's discovery led the Portuguese explorer Vasco da Gama to find the sea route to India in 1497 with the help of a Muslim pilot, Ahmed ibn-Mājid, whom he encountered at Malindi on the east African coast. Mājid showed da Gama his charts, which helped him navigate the Europeans to the Malabar coast, whence da Gama returned to Lisbon in 1499. Before this date Christopher Columbus had sailed to the New World twice and was on his third voyage. But if one had inquired of "informed" Europeans at this time whether the Spanish or the Portuguese voyages were the more important, they would probably have opted for the latter, which promised more immediate gain.

However, the four voyages of Columbus have captured the human imagination to a remarkable degree. The state of European knowledge of the world on the eve of the Iberian discovery of America is suggested by figure 5.4, a reproduction of the gores of the manuscript *erdapfel* (world globe) by Martin Behaim of Nürnberg (Nuremberg), who spent some time in Portugal and who apparently had been engaged in explorations along the African coast.[13] This oldest surviving European terrestrial globe was made in 1492 in Nürnberg and shows the world just prior to the return of Columbus from his first trans-Atlantic voyage. The debt to Ptolemy is obvious and acknowledged, but Behaim includes new information on parts of eastern Asia, resulting particularly from the descriptions of those areas by Marco Polo. The peninsular character of southern Africa is the most striking difference between the work of Behaim and that of Ptolemy. India and Ceylon were remarkably unchanged in comparison with the earlier Ptolemaic delineations of these areas, but they were shortly to assume a more "modern" appearance as the Portuguese went to and from India and points farther east on a regular basis and charted what they found.

Cartographically, Behaim's globe, which was constructed and painted by skilled artisans, is of considerable interest. A globe is, of course, the most accurate means of representing the earth, but it is not the most useful device for all purposes because of the difficulty of measurement and of ascertaining geographical relationships on a sphere. Even when spread out as a series of gores so that the whole world can be seen at one glance, the interruptions (caused in this case by the cuts along poleward-converging meridians) create map-reading problems. Behaim's globe has a diameter of twenty inches and is divided into twelve gores of 30 degrees each; all 360 short degrees are marked where the gores are hinged at the equator. The tropics and the Arctic and Antarctic Circles are shown, but only the 80-degree meridian is drawn from pole to pole. Eurasia, in the Northern Hemisphere's midlatitudes, covers approximately three-quarters of the globe. As a result, the distance between Europe and Asia across the Atlantic is comparatively narrow. A prototype of Behaim's globe—although not this work itself, which he could not have seen—may perhaps have been one of the pieces of evidence that encouraged Columbus, who was a chartmaker as well as a sailor, to venture westward in the hope of finding the coast of Asia. Behaim's *erdapfel* is richly colored: water bodies are blue, except for the Red Sea, which is vermillion on this and many other maps of the period;

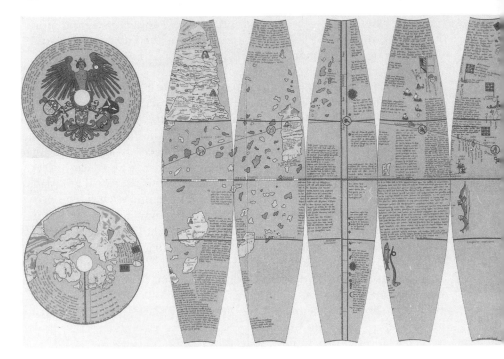

land is generally buff or ochre, with stylized, side-viewed mountains in gray and forests in green. Ships, sea creatures, zodiacal signs, and flags (portolan-style) are also rendered in color. The work was left behind, according to the author, "for the honor and enjoyment of the commonalty of Nürnberg," a rather democratic notion in an autocratic age.

Although printed maps (and globe gores) eventually supplanted hand-drawn examples, there was a long period of overlap, and some of the first important maps of the sea route to India and of the New World are manuscripts. Thus, as early as 1424, the Venetian mapmaker Zuane Pizzigano drew a portolan chart that shows Portuguese discoveries along the coast of West Africa and some mythical Atlantic islands (such as Antillia) that, from later sources, Columbus believed would be "stepping stones" on the way to Asia. Columbus was immediately influenced by a now-lost manuscript world map on an equirectangular (Marinus) projection by Paolo dal Pozzo Toscanelli. This Florentine mathematician and physician sent two copies of the map he had made to Portugal, the later one directly to Columbus, ca. 1480.[14]

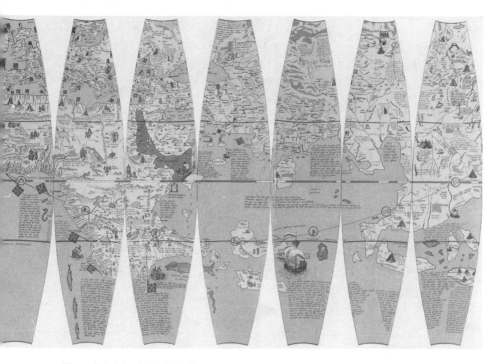

*Figure 5.4.* Martin Behaim's *erdapfel:* gores of the globe and polar caps. Manuscript from Nüremberg (1492).

There is much debate about maps that may have been drawn by Columbus himself, such as a portolan sea chart of the early 1490s, now at Bibliothèque Nationale in Paris, which features a geocentric cosmological diagram. Another manuscript showing a small part of the coast of Hispaniola, formerly attributed to Columbus and alleged to have been made on his first trans-Atlantic voyage, is also of questionable authenticity. However, on his return in 1493, Columbus wrote a letter to his patrons that was translated and printed in several languages at different places, including Barcelona and Rome. A Latin edition published in Basle in 1493 contains maps/illustrations. One of these shows Columbus sailing solo through the islands of the Bahamas, which he renamed. This is an early example of European "acquisition" of territory through cartographic nomenclature, which became a common practice. However, the earliest true general European map of the New World is the portolan-style chart of Juan de la Cosa of 1500, the greatest cartographic treasure of the Museo Naval in Madrid (fig. 5.5). De la Cosa is believed to have been the owner, master, and mate of the *Santa Maria.* It is also thought

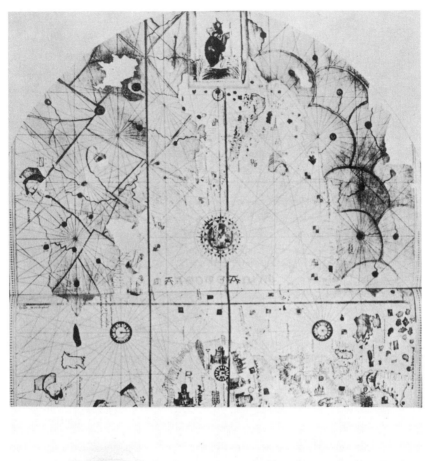

*Figure 5.5.* Juan de la Cosa's map showing New World discoveries of Christopher Columbus and others (ca. 1500): a detail from the manuscript with west orientation and a diagram showing the coverage of the whole map.

that he accompanied Columbus on his first and second New World voyages, but there is some question about this since there may have been two (or more) persons called de la Cosa associated with Columbus. There is also a question about the date of the map: some think that 1500 is too early by about four years, since the map shows areas in South America not discovered by Europeans until that time. The New World is on a different and bigger scale than the Old World (shown on the east side of the de la Cosa map), pointing up the need for a regular map projection to represent such large areas rather than the projectionless portolan chart system. Renaissance projections will be discussed in relation to several maps that were made during the sixteenth century.[15] However, in addition to the discoveries of Columbus, the Juan de la Cosa map shows those of John Cabot for the English (1497) along the North Cape of Asia (as he thought of it), or Newfoundland. Between the representations of the geographical discoveries of Columbus and Cabot on the map there is a vignette of St. Christopher as Christ-bearer. Some have speculated that Columbus was the model for this figure; if so, it is the only life portrait of Columbus now extant.

The conflict between Portugal and Spain for newly discovered lands overseas precipitated a series of papal bulls that after Columbus's first voyage to the New World culminated in the Treaty of Tordesillas in 1494. The treaty's line of demarcation between Spanish claims to the west and Portuguese claims to the east is shown on the portolan-style Cantino map (1502) as 960 nautical miles west of the Cape Verde Islands. Brazil, "discovered" by Pedro Álvars Cabral in 1500, and the "North Cape of Asia," explored by the Côrte Real brothers Gaspar and Miguel from 1500 to 1502, were east of the line and thus claimed by Portugal. The northern coast of South America and the Caribbean islands—explored by Columbus on his first three voyages and by others, including Amerigo Vespucci, before 1500—were claimed by Spain. The line of demarcation was later moved farther west, giving Portugal a larger share of Brazil. The provenance of the Cantino map is interesting: it was drawn by an unknown cartographer in Portugal and smuggled out by Alberto Cantino, who presented it to the Duke of Ferrara (Ercole d'Este) to inform him of the new discoveries. It is now in the Biblioteca Estense at Modena, Italy.

The Juan de la Cosa and the Cantino maps are manuscript, portolan-style charts; the first general *printed* map to show the New World discoveries was made by Giovanni Contarini and his engraver, Francesco Rosselli, in Florence in 1506. Unlike the de la Cosa and Cantino maps, the Contarini map has a regular projection—an orderly system of meridians and parallels, or lines of longitude and latitude. As indicated earlier, Ptolemy is credited with developing projections for geographical purposes, but Renaissance cartographers invented many ingenious solutions to the problem of representing a world rapidly expanding through the discoveries of contemporary explorers (fig. 5.6).

Thus on a fan-shaped projection (fig. 5.6c) Contarini showed most of the Old World, as well as the "North Cape of Asia" discovered by Cabot and the Côrte Reals, immediately to the west of northern Europe,

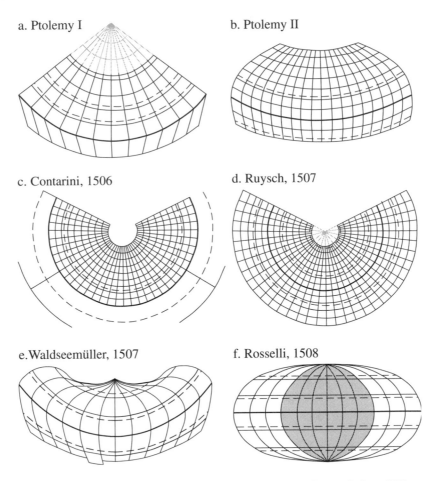

a. Ptolemy I

b. Ptolemy II

c. Contarini, 1506

d. Ruysch, 1507

e. Waldseemüller, 1507

f. Rosselli, 1508

*Figure 5.6.* Projections for world maps used or devised in Europe before 1510. In some cases a selection of the grid has been made for simplicity, and halftoned lines and areas indicate construction.

as mentioned earlier. On this map the discoveries of Columbus and Vespucci in Central and South America are well delineated, but Cipango (Japan) is just west of Cuba, reflecting geographical ideas of the time. The Contarini map, of which only one copy exists, was discovered in 1922 and is now in the British Library.

A map similar to the Contarini map, which is believed to have inspired it, is the engraved map of Johannes Ruysch, a Dutchman living in Germany (fig. 5.6d). Before the discovery of the Contarini map, Ruysch's map was thought to be the earliest printed map of the New World. Ruysch's map accompanied the Rome edition of Ptolemy (1507) and was widely distributed. Both in projection and in map data there are differences between the Contarini and Ruysch maps; one of the most notable is the representation of the North Pole by a curving line on the former and by a point on the latter. Cipango is omitted on Ruysch's map.[16]

There was a gradual realization that a New World—rather than the Indies or Cathay—had been discovered. This is dramatically illustrated on the printed Ptolemaic-type map of Martin Waldeseemüller (1507), who worked at Saint-Dié in the Rhineland, showing two separated parts of the New World, north and south (figs. 5.7 and 5.6e). This map, one of the most important in the history of cartography, is very large (fifty-three by ninety-four inches in six sheets) and difficult to reproduce.[17] Only one copy of this woodcut map exists, at Schloss Wolfegg, Wüttenberg, Germany. It is the first dated map on which the name America appears—in honor of Amerigo Vespucci—in South America. Waldeseemüller was influenced by Vespucci's accounts (including the so-called Soderini letter) of his 1499 voyage to the north coast of South America, *Novus Orbis*. Vespucci deserves credit for appreciating in a way that Columbus never did that a fourth continent had been added to those of Europe, Africa, and Asia. Later, Waldeseemüller realized that he had given too much credit for the discovery of the New World to Vespucci. He tried to correct this on a plane chart of 1513, but it was too late— the name America had stuck. The inscription in South America on the 1513 map can be translated as "This land and adjacent islands were discovered by Columbus of Genoa for the monarchs of Castile." On Waldeseemüller's world map the Americas are represented as of very limited longitudinal extent because they are fitted into a modified Ptolemaic framework and because the west coasts had not yet been explored. On the top of the map are two globular hemispheric projections with representations of Ptolemy as the cartographer of the Old World and Vespucci as the cartographer of the New.

CHOR

ZEPHIR

AEOLVS

UNIVERSALIS          COSMOGRAPHIA          SECVNDVM PTHOLOMÆI TRA          DITIONE

*Figure 5.7.* Martin Waldseemüller's world map (1507), showing the Americas to be of a very limited longitudinal extent and the name America (in South America) for the first time on a printed map.

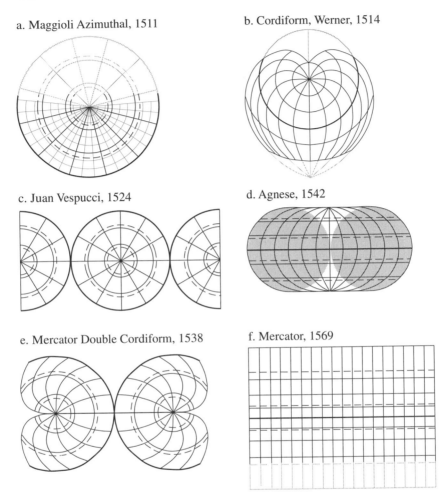

a. Maggioli Azimuthal, 1511

b. Cordiform, Werner, 1514

c. Juan Vespucci, 1524

d. Agnese, 1542

e. Mercator Double Cordiform, 1538

f. Mercator, 1569

*Figure 5.8.* Projections for world maps devised in Europe, 1511–1569.
In some cases a selection of the grid has been made for simplicity,
and halftoned lines and areas indicate construction.

In Waldeseemüller's 1507 map we can see the beginning of the evolution of the cordiform (heart-shaped) projection (fig. 5.8b), which was further developed by Bernard Sylvanus (1511), Johannes Werner (1514), and Oronce Fine (1521), among others.[18] Another ingenious way of representing the expanding world of the Europeans was by means of a printed map on an oval projection by Francesco Rosselli (1508), now at the National Maritime Museum, Greenwich, England (fig. 5.6f).

A manuscript map of 1511 by Vesconte de Maggioli, a Genoese working in Naples, is on a new, polar azimuthal projection (fig. 5.8a), with the Old and New Worlds (Siberia and the "North Cape of Asia") connected across the North Pole. Other civilizations soon learned of the European discoveries, and America was shortly delineated on Islamic maps (such as the Piri Reis map, by a Turkish corsair and admiral of this name, from 1513) and, somewhat later, on Chinese maps.[19] Meanwhile the process of globalization proceeded apace, with new discoveries being made and new projections devised.

The making of celestial and terrestrial globes, some of the latter used even for high-latitude navigation, became an important activity in the workshops of European instrument makers in the Renaissance, as it had been earlier in Asia. The first "professional" globe maker in Europe was Johann Schöner, whose terrestrial globe of 1520 shows South America and Africa with surprising accuracy, although southern Asia is not as well represented.

In 1508, Vespucci became the first pilot-major, or supervisor of maps and charts, of the Casa de la Contratación de las Indias in Seville. The responsibilities of the institution included the entering of new information on a master map (the *Padrón General*, or official record of discoveries); the supervision of charts and instruments carried by seamen; and the examination of pilots. Although the *Padrón* is lost, it is believed to be approximated by the world chart of Diego Ribero (1529). On Ribero's map, the east coasts of the Americas are well delineated, based on the explorations of Columbus, Cabot, Giovanni da Verrazzano, and others.[20] The west coasts of both continents are not drawn because they were not yet explored, except by Vasco Núñez de Balboa (1513) and his successors in Central America. Ribero was a Portuguese working for Charles V (Carlos I of Spain), the Holy Roman Emperor.

The circumnavigation of the globe by Ferdinand Magellan and Juan Sebastián de Elcano between 1519 and 1522 precipitated much original cartography. A good deal of this has to do with the Far East and the extension of the line of demarcation between Spanish and Portuguese claims in the East Indies. One such map, an elaboration of the polar azimuthal projection of Maggioli, is Juan Vespucci's double hemispheric projection. Juan Vespucci was the nephew and heir of Amerigo and, like his uncle, was a cartographer and examiner of pilots in the Spanish service. The map was drawn in 1524 in an attempt to settle the dispute between Spanish and Portuguese claims in the Moluccas (the Southeast Asian Spice Islands). The projection is novel (fig. 5.8c), showing the Northern and Southern Hemispheres, interrupted, with the equator and tropical and Arctic and Antarctic circles indicated and named.

Charles V commissioned a map for his son, later Philip II of Spain, showing the track of the circumnavigators Magellan and de Elcano. It was developed on an ovoid projection, on which the poles are represented as half the length of the equator (fig. 5.8d), suggestive of the modern Eckert III projection. Of special interest on this map, made in 1542, is the delineation of the Americas with the Strait of Magellan, the Colorado River, and Baja California. The map is the work of the Venetian cartographer Battista Agnese, who received information on the explorations of Hernán Cortés and his lieutenants on the west coast of Mexico after 1530. However, as indicated, the delineation of the track of Magellan across the Pacific (1519–21) to the Philippines, where he was killed, and the return of the *Victoria (Vittoria)* via the Cape of Good Hope under de Elcano (1521–22) was the main purpose of the map.

The hallmark of the Renaissance was the universal genius, and we find that a number of workers who are thought of primarily in other connections were engaged in cartography, including Leonardo da Vinci, who drew urban and engineering plans.[21] Among cartographers considered more particularly as artists were Albrecht Dürer, the inventor of etching (not much used in cartographic reproduction), and Hans Holbein the Younger, who drew a world map on an oval projection that was printed in Basle in 1532.[22] Those more specifically associated with mapping include Giovanni Contarini, whose world map was discussed earlier; Peter Bienewitz (Peter Apian); and Gemma Frisius. Frisius, an astronomer and mathematician (credited with the development of surveying by triangulation) as well as a cartographer, was the teacher of Gerardus Mercator (Gerhard Kremer). Mercator was born in Flanders in 1512, and while at the University of Louvain he became a student of Frisius, for whom he engraved a globe (ca. 1536). In 1538 he went on to publish a double cordiform projection, an idea he obtained from Fine, on which he applied the name Americae to the continent north of the Caribbean, as the landmass south of this sea had already been known for over thirty years (fig. 5.8e). Following Frisius, Mercator also engaged in land surveying, producing highly accurate maps of Europe in 1541 and 1554, the latter being on a conical projection with two standard parallels. Mercator was an expert engraver and introduced italic lettering to northern Europe.[23] Apparently unaware of the much earlier work of Muslim astronomers on this subject (referred to in the preceding chapter), but utilizing the best available itineraries and charts as source materials, Mercator reduced the map length of the Mediterranean from Ptolemy's figure of 62 degrees to 52 degrees, which, though still too long by 10 degrees, nevertheless represented a great cartographic improvement. Since antiquity, latitude had been measured with considerable accuracy, but longitude, especially on shipboard, was difficult to determine before the invention of reliable, portable timepieces (chronometers) in the second half of the eighteenth century.[24]

All of the previous accomplishments of Mercator, who was by this time established in Duisburg in the Rhineland, were eclipsed by the publication in 1569 of a great world map on the projection that bears his name (figs. 5.8f and 5.9). It should be mentioned here that there are indications that, before Mercator, portolan and Iberian plane chartmakers were struggling toward a solution for representing loxodromes (rhumb lines) on maps. In fact, Mercator was anticipated in this projection by Erhard Etzlaub, a Nürnberg instrument maker who, among other contributions, engraved two maps on compass lids in 1511 and 1513. On these, gradations of parallels from the equator (0 degrees) to 64 degrees north latitude are in "increasing degrees," and the maps cover about 65 degrees longitude, which is not expressed. Coastlines include North Africa to northern Europe, and the maps, like all surviving examples by this cartographer, are south-oriented. Eztlaub's projection extends over about only one-tenth of the surface of the globe, and it remained for Mercator to construct a world map using this principle.

Mercator's projection is a superb example of the positive value of a map over a globe for a specific purpose. Like several other projections, the Mercator is conformal (shapes around a point are correct), but it also has a unique property: straight lines are rhumb lines or loxodromes (lines of constant compass bearing). This quality, which makes the projection of great value to the navigator, is accomplished by increasing the spacing of the parallels by specified amounts from equator to poles. Although Mercator was the first cartographer to produce a true world navigational chart with graticules on which a compass line intersects each meridian at a constant, given angle, he left no instructions for its construction and utilization. Apparently Mercator derived his projection empirically, and it was the English mathematician Edward Wright who provided an analysis of its properties, which he published in *Certaine Errors in Navigation* (1599). An English world map of this same date on the Mercator projection by Wright and Emery Molyneux was bound with Richard Hakluyt's *Voyages*. It shows the results of the circumnavigation of Francis Drake (1578–80) and is referred to by Shakespeare in *Twelfth Night* (Act III, Scene 2): "He does smile his face into more lynes than are in the new Mappe with the augmentation of the Indies."[25] Unfortunately, over the years the Mercator projection has been used to show earth distributions for which it is totally unsuited because of extreme

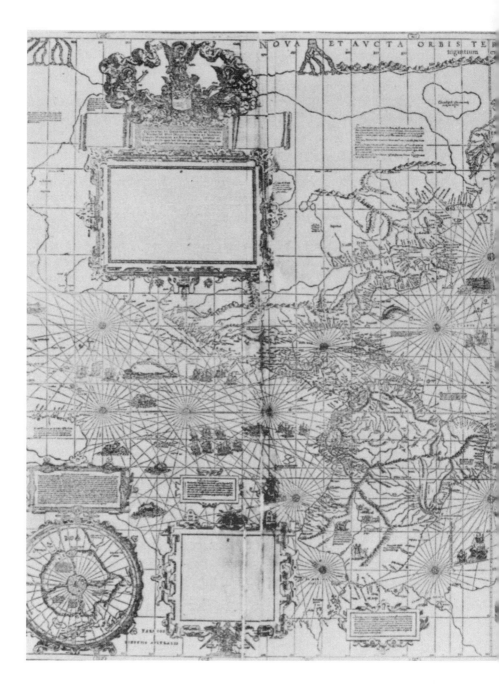

*Figure 5.9.* World map (engraving) by Gerardus Mercator on the projection that bears his name (1569).

deformation (distortion) in the mid- and higher latitudes. Furthermore, the highest latitudes cannot be shown at all on the equatorial case of the Mercator projection.

We have been concerned with the projection used by Mercator in his map of 1569; we should now briefly consider the geographical content of that chart. Between the publication of Waldseemüller's map of 1507 and Mercator's map of some sixty years later, great progress had been made in the European exploration of the world's coastlands. As indicated above, in 1513 Balboa sighted the Great South Sea from a peak in Darien. Seven years later, Magellan sailed into the Pacific, to be followed by a dozen major Iberian expeditions before the appearance of Mercator's map in 1569. This chart reflects much of that progress in the delineation of the west coasts of South and Central America. Although a decorative cartouche blocks much of North America, the great longitudinal extent of this continent is suggested. Baja California is firmly attached to the continent, though it was later represented as an island.[26] South and Southeast Asia are more convincingly represented on Mercator's map than on Waldseemüller's, as a result of more than fifty years of Portuguese exploration in that area. There is also an outline of a great continent in the South Seas, *Terra Australis,* recalling Crates' concept. This was to be a fata morgana that persisted until finally disproved by the discoveries of Captain Cook in the second half of the eighteenth century.

For a quarter of a century after the publication of his world chart in 1569, Mercator's cartographic activities continued; he was working on a great atlas at the time of his death in 1594. In the following year, the first edition of the work was published under the direction of his son Rumold as *Atlas sive cosmographicae meditationes de fabrica mundi et fabricati figura.*[27] The Mercator family continued to publish the *Atlas* until the plates were bought by Jodocus Hondius in 1604. Prior to this time Hondius was in England, where he made the gores for the Molyneux globe, the first to be printed in that country. When he returned to Amsterdam, Hondius "scooped" Wright by publishing the latter's explication of the Mercator projection in a text printed in 1597. After acquiring the plates of Mercator's *Atlas,* Hondius published an edition with additional maps in 1606 and translated editions in French (1607), German (1609), and Dutch (1621). Following Hondius's death in 1612, the *Atlas* was published by his widow and their sons. When Johannes Jansson, a publisher, married Hondius's daughter Elizabeth, he was brought into the family business, so that by the time of the English edition of 1636, the *Atlas* bore three names: Mercator–Hondius–Janssonius. Hondius and his successors also published sheet maps, including the broadside map (1593) showing the track of the circumnavigation of the world by Drake on a double hemi-spheric, stereographic projection, with inset views.[28]

The appearance of the first edition of Mercator's *Atlas* was antici-pated by some twenty-five years by the *Theatrum orbis terrarum* of Abra-ham Ortelius (Oertel) of Antwerp. Ortelius, the friend and rival of Mer-cator, is credited with producing the first modern uniformly bound collection of printed maps designed especially for this type of publica-tion. Previously sheet maps had been assembled in atlas form, as in the various editions of Ptolemy's *Geographia,* and in bound collections of mis-cellaneous maps by Antonio Lafreri and others in Italy. Lafreri (Antoine du Pérac Lafréry) was a French engraver who in the middle of the six-teenth century settled in Rome, where he assembled and sold bound collections of maps *(atlas factice)* from various sources. The title page of his collections featured the mythological figure of Atlas carrying the world on his shoulders, but it is undoubtedly because Mercator later employed the term *atlas* for a book of maps that it is in use today. The first edition of Ortelius's *Theatrum,* containing seventy maps, appeared in May 1570; it was immediately successful and was followed by two more editions in the same year. Ortelius employed the compilations of many cartographers (usually one per country with acknowledgment) but had the maps engraved on a uniform format.

To illustrate this cartographic form, we have reproduced the map of England and Wales from Ortelius's *Theatrum* (fig. 5.10). The original map occupies two leaves with an image size of fifteen by eighteen and a quarter inches and is of intermediate or medium scale. In this respect, it stands in contrast to maps of small or geographical scale (such as fig. 5.9) and those of large, or topographical, scale, which will be considered later. In its several elements, figure 5.10 is representative of much Re-naissance cartography. The title of the map and the name of the author, Humphrey Lhuyd, a Welsh physician who practiced in London, are con-tained in an ornate cartouche or inset box. This feature, with its escutch-eon, supporters, and so forth, is very typical of the cartography of the time. A bar or graphical scale is in another small cartouche, and in the sea are various ships of the period and two sea creatures. The lettering—particularly the italic lettering, with sweeps and swash lines in the water bodies—is decorative and suitable to the style of the map. On the land areas are several prominent symbols: profile "molehills" or "sugar loaves" representing hills and mountains, with no real differentiation between higher and lower ranges and no landforms other than uplands and plains represented; scattered buildings in groups, also in profile, representing urban settlements, with little differentiation between larger and smaller places; and rivers widening from source to mouth, often

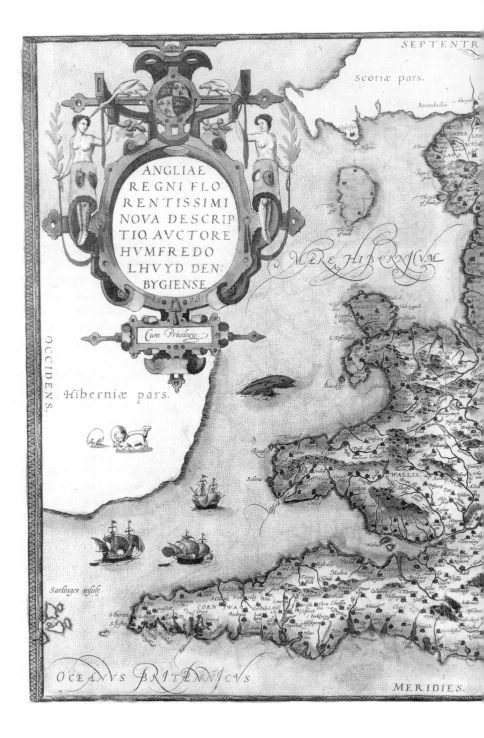

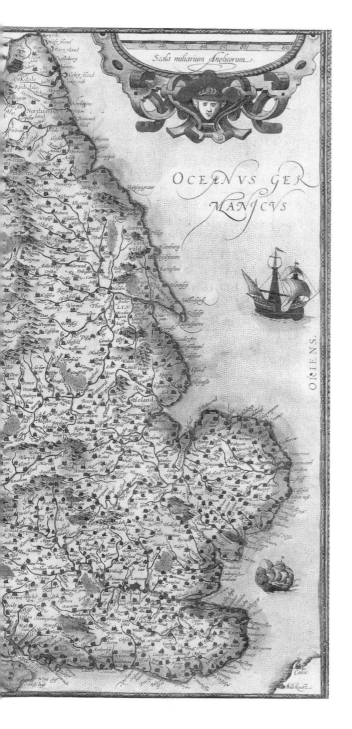

*Figure 5.10.* Engraved plate from Abraham Ortelius's *Theatrum orbis terrarum* (1570) showing England and Wales, with information supplied by Humphrey Lhuyd; the original is hand-colored.

with exaggerated estuaries. No latitude and longitude indications appear, although there are grids on smaller-scale maps in the *Theatrum*. The tonal quality of figure 5.10 resulting from the subsequent hand coloring of the engraving is, in this instance, tastefully executed, but it often detracts from the quality of the work. The stippling in the sea is on the original plate, and the coastlines produce a recognizable outline of the country but, understandably, lack the rigor of a modern controlled survey map. Coastlines are everywhere emphasized by horizontal shading, but areas that are not the prime subject of the map (parts of Ireland, Scotland, and France, in this case) usually have only major cities and landmarks identified; these areas are detailed in separate maps in the atlas. Orientation is indicated on the four sides of the map with the names of the cardinal directions. The whole is enclosed with a border that gives the effect of a molding or picture frame; indeed, many such maps have been removed from atlases and are now used as pictures or as decorative elements on lampshades and coffee tables, leading to the destruction of the atlases of which they were once a part. On the reverse side of the hinged sheet is a description of the area mapped; such descriptions typically accompany these works.

Between its appearance in 1570 and the last printing, dated 1612, the *Theatrum* was published in more than forty editions and translated into Dutch, German, French, Spanish, Italian, and English. This period in cartographic history has been called the age of atlases because the work of Ortelius and Mercator inspired many others to engage in this lucrative trade, Hondius and Jansson (as indicated) and Blaeu and Visscher being among the best known.[29] A number of these cartographers were also makers and publishers of other cartographic products: globes, wall maps, and sea charts. The first true atlas of printed sea charts was *De Spieghel der Zeevaerdt* by Lucas Janszoon Waghenaer (Leiden, 1584).[30] This publication owed a great deal to a tradition of manuscript sea charts in which Italian, Dieppese, and earlier Netherlandish work was particularly important. In addition to charts with soundings in fathoms and conventional signs, *De Spieghel* contains other hydrographic information, including coastal views. It proved so popular that a Latin edition was published in 1586 and an English translation was made by Sir Anthony Ashley as the *Mariner's Mirrour* in 1588, the Armada year. A reproduction of the chart of the coasts of Western Europe from the first English edition of Waghenaer is used to illustrate this cartographic genre (fig. 5.11). The inheritance from the portolan chart in the compass roses and rhumb lines is obvious. At first the charts depicted only the European coasts, often in considerable detail, but gradually the whole world was covered. This was the case with Robert Dudley's *Arcano del mare*, compiled and published in Italy in 1664, the first sea atlas to use the Mercator projection throughout. The trade in manuscript sea charts flourished side by side with the printed chart business, some practitioners being involved in both enterprises, until the beginning of the eighteenth century. Throughout this period the dominance of the Dutch in hydrographic mapping was so great that sea atlases were known as *waggoners* (after Waghenaer) or variants of this term in other languages.

Cartography developed in yet other directions in the Renaissance. Ortelius had included the *Parergon,* a supplement of historical maps both biblical and classical, in the 1579 edition of the *Theatrum,* the first edition printed by the famous Plantin Press. Between 1572 and 1618, an atlas of city plans titled *Civitates orbis terrarum* was published in Cologne by Georg Braun (Joris Bruin) and Frans Hogenberg.[31] Hogenberg had been employed by Ortelius as an engraver, and in conception, but not in subject, the *Civitates* resembles the *Theatrum.* An example of the contents of the Braun and Hogenberg work is included as figure 5.12. This oblique or three-quarter (bird's-eye) view of Bruges shows a city that has undergone little change in its basic morphology since the engraving was made in the sixteenth century. Other cities represented (such as London, which was devastated by the Great Fire of 1666) have since changed almost beyond recognition. This cartographic form was also imitated but not materially improved upon for many years. In our present age, with its emphasis on urban affairs, sources such as the *Civitates*—which contains views of Mexico City and Cuzco as well as cities of Europe, Africa, and Asia—are of particular interest.

European cartographic production in its various expressions in the late fifteenth and the sixteenth centuries was focused in the valley of the Rhine and along its distributaries: from Waldseemüller in Saint-Dié; to Mercator in Duisburg, where he moved after 1552 from the Netherlands; to Ortelius in Antwerp; to Hondius, Janssonius, Blaeu, Visscher, Van Keulen, and others in Amsterdam. Another leading Rhineland cartographer was Sebastian Münster of Basle, who made the first separate maps of the continents, including America (1540). But other areas and directions were involved in this pan-European enterprise. The importance of Italy, especially Florence and Venice, has been discussed in connection with the early compilation and printing of maps. An outstanding Venetian cartographer of the Renaissance was Giacomo Gastaldi (1500–66), on whose world map of 1546 many place-names in America appear for the first time. We have already mentioned the Nürnberg instrument

*Figure 5.11.* Chart of Western Europe from *The Mariner's Mirrour* (London, 1588), a translation of *De Spieghel der Zeevaerdt* by Lucas Janszoon Waghenaer (1584), edited by Sir Anthony Ashley.

BRVGÆ, *vulgo Brugk: Teuto:*
*niæ Flandriæ vrbs omnium*
*pulcherrima, nitidiſſimaque, publi*
*carum ſiquidem, priuatarumque*
*ædium in hac vrbe ſplendor et*
*magnificentia, omnem ratio:*
*nem, omnem dicendi faculta:*
*tem ſuperat. Optimam vrbi:*
*um formam, hoc eſt, orbicula:*
*rem, ſitu obtinet, aquis pro:*
*bè inſtructa, duplici foſſa*
*ambitur; florentiſſimum quõ*
*dam emporium fuit.*

*Figure 5.12.* The city of Bruges from *Civitates orbis terrarum* by Georg Braun and Frans Hogenberg (1572).

maker, physician, and cartographer Etzlaub and his proto-Mercator chart. He also made a world map, now lost, as well as one covering a wide swath across Europe from Denmark to Rome and another of the environs of his home city, on both of which he plotted routes and distances in German miles.[32] Working in Zurich at about this time (late fifteenth century) was the Swiss physician Konrad Türst, who drew a south-oriented map of his homeland with its lakes, mountains, and cities, the last of these as thumbnail sketches, reflecting his training as an artist.[33] A great school of manuscript chartmakers arose in the northern French seaport of Dieppe, from which one Jean Rotz defected to England in the mid–sixteenth century, taking with him a manuscript sea atlas he had made, dedicated to Henry VIII.[34] This king invited Italian engineers to assist him in designing and constructing coastal fortifications, but the English were to make considerable contributions to these enterprises, including the drawing of plans, during the Tudor period. Robert Thorne designed a world map in 1527 (first printed in 1582) to persuade the English to seek a Northeast or Northwest Passage to the Spice Islands; on this map, political divisions are delimited and named. Especially important were the maps of all the counties of England by Christopher Saxton (ca. 1542–1606), as well as his large map of the whole of the kingdom.[35] Saxton's work grew out of a land-estate surveying tradition that extended back to the late Middle Ages, leading to maps known in England as *terriers*. Saxton's county maps, which were derived from instrumental surveys, were soon emulated and became the basis of such cartography for a century and a half.

Local mapping—whether using relatively sophisticated instruments and materials or sketches by indigenous peoples whom the Europeans encountered—could be incorporated into global map projections that were devised in the European Renaissance but had their origins in cosmological and geodetic concepts of classical antiquity. The printed world map on a systematic projection was perhaps the most distinctive contribution to cartography of the period and the area.

# SIX

## Cartography in the Scientific Revolution and the Enlightenment

The dissemination of knowledge made possible by the invention of printing in Europe in the last decades of the fifteenth century was an important factor in the spectacular increase in scientific activity that followed. With the publication of *De revolutionibus orbium coelestium* in 1543, Nicholas Copernicus (Niklas Koppernigk, Mikołaj Kopernik), who realized that the earth is a planet, revived the heliocentric theory of the universe proposed centuries earlier by Aristarchus. Copernicus (1473–1543), a cartographer as well as an astronomer, produced maps of Prussia (1529) and of Lithuania, now lost.[1]

Although a greater theorist than an observer, Copernicus ushered in a period in which, more than ever before, experimentation was related to observation.[2] A concomitant development was the invention of new instruments and the improvement of existing ones. This not only led to an increase in accuracy but also extended the range of observation. Progress was made on a number of fronts, some of which affected cartography, directly or indirectly.

As indicated before, the resulting maps are of more concern here than the methods used in their construction, but a few important technological milestones in mapping will be discussed. Triangulation, the fixing of places by intersecting rays, was described by Gemma Frisius in 1533; the plane table, with a sighting rule on the drawing surface, which enabled the map to be made at the same time that the angles were drawn, was reported by Leonard Digges in 1571; various tables were compiled, including that for ephemerides by Regiomontanus, Johannes Müller (1436–76), and his student-patron, Bernhard Walther (1430–1504), and that for logarithms by John Napier (1550–1617) and Henry Briggs (1561–1630); the pendulum clock, allowing more accurate determination of longitude at fixed points of observation, proposed by Galileo, was built by Christian Huygens (1629–95) in 1657. In addition, the seventeenth- and eighteenth-century surveyor had available, from the previous century and much earlier, the odometer and the magnetic compass. A derivative of the latter instrument, the theodolite, was improved toward the end of the period to enable horizontal and vertical angles to be measured simultaneously (altazimuth theodolite).

Although remembered primarily for his work in physics and astronomy, Galileo Galilei (1564–1642) also made maps. Galileo learned of the development of telescopic lenses in the Netherlands and, in 1609, constructed telescopes himself. He was the first scientist to employ the telescope for research purposes and, with a three-power instrument, made what are presumably the first lunar charts by this means. Galileo's original maps were destroyed, along with many of his other works, but several engravings of his lunar drawings are found in his *Sidereus nuncius* (1610).[3] The lunar sketch map (fig. 6.1), though crude, was the first to show craters, which Galileo attempted to measure, and seas *(maria)*, which he shortly realized were not water bodies. The fact that as revealed by Galileo a celestial body such as the moon was a less than ideal form had most important theological implications. Through a "remote sensing" device, Galileo first mapped a surface that humans would not explore directly for more than 350 years. Galileo discovered two of the moon's librations and initiated the serious study of selenography through his maps and writings.

The work and life of Galileo most dramatically illustrate the spirit of the Scientific Revolution, but later scientists contributed more to cartography specifically. After Galileo, a generally more favorable attitude toward science existed in Europe, particularly north of the Alps. Following the lead of Italy in this regard, societies were founded in France (Académie Royale des Sciences) and England (Royal Society of London) in the mid–seventeenth century to foster scientific inquiry. Under this impetus, and independently, we can recognize several directions in which cartography moved in the seventeenth and eighteenth centuries, including cadastral, route, thematic, hydrographic, and topographic mapping. All of these traditions have been recognized earlier, and all were to be advanced in later centuries, as will be indicated subsequently. But with the greater interest in physical science in the period presently under discussion, cartography, along with other related activities, experienced profound changes that have continued and expanded up to the present time.

In the wake of increasing nationalism and colonialism in the seventeenth and eighteenth centuries, cadastral mapping was an instrument used for economic and political purposes. Many examples could be given, but the case of Ireland is instructive. An early attempt by the En-

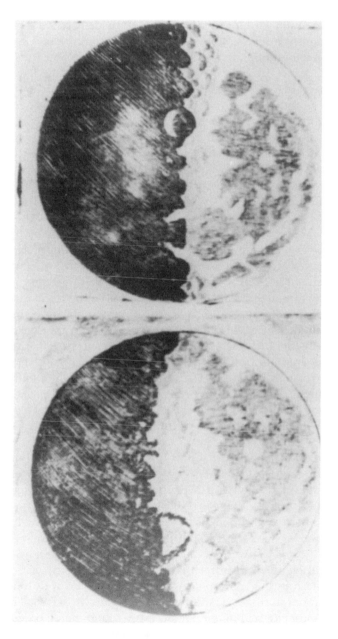

*Figure 6.1.* Engraving of a lunar sketch map by
Galileo Galilei from his *Sidereus nuncius* (1610).

glish to organize planned rural settlements in Ireland was the Munster Plantation (1585–86), for which maps were made of a country regarded by the English as suitable for "planting," or colonization. A much more important and durable cadastral mapping enterprise was the Down Survey, which was conducted by the political economist Dr. (later Sir) William Petty (1623–87) as surveyor general of Ireland (1655–56) under the Protectorate. Following the English Civil War, in which Ireland was vitally involved, Petty surveyed forfeited lands that were "laid down" on maps of parishes at a scale of three or six inches to the mile. In addition to boundaries, cultivable land, woods, bogs, mountains, and, in some cases, roads are shown, presaging land-use maps of a much later date. The surveys were subsequently organized into a series of county or regional maps, which were engraved in Amsterdam and published in 1685, two years before Petty's death. As a physician and one of the founders of the Royal Society, Petty had his hand on the pulse of scientific activity in his time and made notable contributions to the study of statistics, for which he is now chiefly remembered.[4]

We have noticed that road maps have been an important, if intermittent, cartographic form since at least the time of the Roman itineraries. After the fall of Rome, the great system of highways connecting the empire lapsed into disrepair because there was no centralized organization to maintain them. In the later Middle Ages, routes were reopened for military operations and for pilgrimages and occasionally were the subject of maps, such as Matthew Paris's map from London to southern Italy and Erhard Etzlaub's *Rom Weg*, referred to earlier. Later, roads were often omitted on maps, as in the case of Saxton's county maps, but main roads were included in works in the same field by his successor, John Norden (1548–1625), who was also a pioneer in preparing triangular distance tables. Even more innovative was the work of John Ogilby (1600–76), a Scot, who was Master of the King's Revels, bookseller, translator, printer, and finally cartographer. In these last two capacities, toward the end of his life, he published *Britannia* (London, 1675), in which by a series of strip maps of the kingdom, he gives the distance in miles between places along major post highways, from London to main provincial towns.[5] This procedure was followed by others in the seventeenth and eighteenth centuries and later, both in strip and in "areal" maps. A derivative is represented by the distance maps of John Adams, a topographer and lawyer, who approached the Royal Society in 1681 with a proposal for mapping England using triangulation based on astronomical measurements. This was not funded but, as we shall see, it was being accomplished at this time in France, which had become the great rival of England for scientific, political, and colonial hegemony. Meanwhile the Dutch continued to dominate map and chart publishing but were often content to reprint plates with little or no amendment. Out-of-date plates from the Low Countries were used by Moses Pitt for the abortive *English Atlas* published at Oxford (1680–83).[6] We can now turn to thematic maps, perhaps the type most neglected by historians of cartography until quite recently.

A *thematic map* is designed to serve some special purpose or to illustrate a particular subject, in contrast to a general map, on which a variety of phenomena (landforms, lines of transportation, settlements, political boundaries, and so forth) appear together. The distinction between general and thematic maps is not altogether sharp, but the latter type uses coastlines, boundaries, and places (base data) only as points of reference for the phenomenon being mapped (map data, or theme of the map) and not for their own sake. One of the most significant contributors to thematic mapping was the English astronomer Edmond Halley (1656–1742), best known for his prediction of the periodic return of the comet that bears his name. Of course Halley was not the first to make thematic maps.[7] We have mentioned the biblical and historical maps in Ortelius's *Theatrum;* we know that Oronce Fine, the well-known French cartographer, made biblical maps (now lost) in the mid–sixteenth century; and other examples could be cited. But these are very different from the thematic maps of Halley, who illustrated a number of his own scientific theories by cartographical means.

Edmond Halley, whose mapping endeavors well exemplify the cartography of the Scientific Revolution, was a fellow and, for a time, clerk to the two secretaries of the then-youthful Royal Society of London. Through this connection he became acquainted with many of the greatest scientists of his age, including Johannes Hevelius of Danzig (1611–87), the foremost lunar cartographer of the seventeenth century; Giovanni Domenico (later Jean-Dominique) Cassini (1625–1712), who supervised the construction of an earth map *(planisphère terrestre)* on the floor of the Paris Observatory; and Sir Isaac Newton (1642–1727), who contributed specifically to cartography and geodesy through his announcement, before it was proved, that the earth is an oblate (polar-flattened) spheroid rather than a prolate (equatorial-flattened) or a perfectly spherical one, as supported by the French scientific establishment at the time.[8]

Halley's first significant cartographic venture was a celestial planisphere (star chart) of the constellations of the Southern Hemisphere, made during a stay of about a year at St. Helena and published in 1678. Some years after this, Halley became editor of the *Philosophical Transactions,* in which he published his first important terrestrial map in 1686

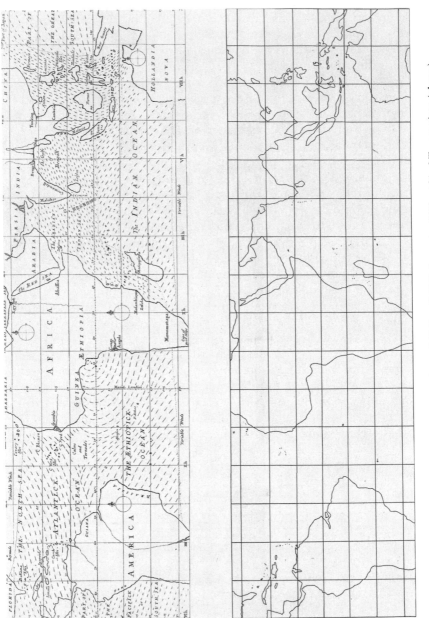

*Figure 6.2.* Chart of the trade winds by Edmond Halley (1686) from the *Philosophical Transactions* (above).
Below, for comparison, is a map on the same scale and projection to indicate a modern delineation of the world's coastlines.

(fig. 6.2). This map, which accompanies an article by Halley on the trade winds, has been called the first meteorological chart.[9] It well illustrates a new trend in cartography, focusing, as it does, on a single physical theme—the direction of the prevailing winds in the lower latitudes. Halley selected a suitable projection (the Mercator) for showing the phenomenon being mapped (the map data).[10] The grid is composed of 10-degree lines of latitude and 15-degree lines of longitude (based on the prime meridian of London), each of which represents one hour of earth rotation. No extraneous decoration appears on this chart. The prevailing winds are shown generally by tapering strokes, the tails indicating the direction from which the wind usually comes. Halley used arrows, which are now the conventional symbols for such phenomena, only in the Cape Verde area. A minimum of base data needed for showing the thematic distribution appears on the map; some may even object to its being so designated since it lacks a title and scale indication. The addition of the north and west coast of Australia to the world map resulted largely from the sixteenth- and seventeenth-century explorations of the Portuguese, Spanish, and Dutch in this area. The map below Halley's wind chart in figure 6.2 shows coastlines from a present-day source drawn on the same projection for comparison.

In 1698 Halley was given a temporary commission as a captain in the Royal Navy for the purpose of determining longitude of places accurately and of investigating the earth's magnetic field. He took command of a small ship, the *Paramore,* and embarked on a voyage sponsored by the Royal Society, which has been called "the first sea journey undertaken for a purely scientific object."[11] On this voyage—really two voyages—Halley made about 150 observations of magnetic declination over the Atlantic, from approximately 50 degrees north latitude to 50 degrees south latitude, which were obtained under the most difficult circumstances and were basic to his research. Shortly after his return to England in 1700, Halley published a map that he titled "A New and Correct Chart Shewing the Variation of the Compass in the Western and Southern Ocean" (fig. 6.3). The map uses isogones (lines of equal magnetic declination, or variation in degrees from geographical north) and is the first such printed map extant, as well as the first published isoline map of any kind.[12] In thematic, quantitative cartography, isolines (isarithms) are lines connecting points with equal intensity of phenomena that have transitional degrees of intensity. Thus coastlines, which normally and theoretically depict the same intensity, cannot be considered isolines in this sense (an exception, of course, would be when high and low tide lines are shown on the same map). We shall encounter this fundamental

# Defcription
### AND
## USES
Of a New and Correct
### SEA-CHART
Of the Weftern and Southern
## OCEAN,
#### Shewing the Variations of the
#### COMPASS.

THE Projection of this *Chart* is what is commonly called *Mercator*'s; but from its particular Use in *Navigation*, ought rather to be named the *Nautical*; as being the only true and fufficient *CHART* for the *Sea*. It is fuppofed, that all fuch as take Charge of Ships in long Voyages, are fo far acquainted with its Ufe, as not to need any Directions here. I fhall only take the Liberty to affure the Reader, that having taken all poffible Care, as well from Aftronomical Obfervations, as Journals, to afcertain the Scituation and Form of this *Chart*, as to its principal Parts, and the Dimenfions of the feveral Oceans; he is not to expect that we fhould defcend to all the Particularities neceffary for the Coafter, our Scale not permitting it. What is here properly New, is the *Curve-Lines* drawn over the feveral Seas, to fhew the Degrees of the *Variation* of the *Magnetical Needle*, or *Sea Compafs*: which are defign'd according to what I my felf found in the *Weftern* and *Southern* Oceans, in a Voyage I purpofely made at the Publick Charge in the Year of our Lord 1700.

That this may be the better underftood, the curious Mariner is defired to obferve, that in this *Chart* the Double Line paffing near *Bermudas*, the *Cape Verde Ifles*, and *Saint Helena* every where divides the *Eaft* and *Weft* Variation in this Ocean, and that on the whole Coaft of *Europe* and *Africa* the Variation is Wefterly, as on the more Northerly Coafts of *America*, but on the more Southerly Parts of *America* 'tis Eafterly. The Degrees of *Variation*, or how much the Compafs Needle declines from the true North on either fide is reckoned by the Number of the Lines on each fide the double Curve, which I call the *Line of No Variation*; on each fifth and tenth is diftinguifhed in its Streak, and numbered accordingly, fo that in what Place foever your fhip is, you find the *Variation* by Infpection.

That this may be the fuller underftood, take thefe Examples. At *Madera* the *Variation* is 3 and ¼d. Weft; at *Barbadoes* 5½d. Eaft; at *Antakes* 7d. Weft; at Cape *Race* in *Newfoundland* 14d. Weft; at the Mouth of *Rio de Plata* : 8d. Eaft, &c. And this may fuffice by way of Defcription.

As to the Ufes of this *Chart*, they will eafily be underftood, efpecially by fuch as are acquainted with the Azimuth Compafs, to be, to correct the Courfe of Ships at Sea: For if the Variation of the Compafs be not allowed, all Reckonings muft be fo far erroneous: And in continued Cloudy Weather, or where the Mariner is not provided to obferve this Variation duly, the *Chart* will readily fhew him what Allowances

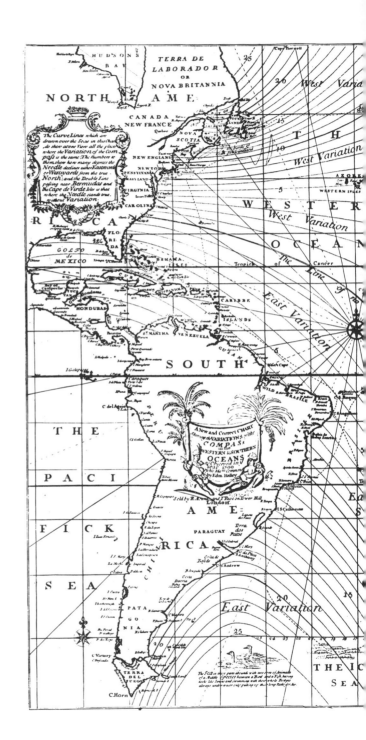

he muſt make for this Default of his Compaſs, and thereby rectify his Journal.

But this Correction of the Courſe is in no caſe ſo neceſſary as in running down on a Parallel *Eaſt* or *Weſt* to hit a *Port*: For if being in your Latitude at the Diſtance of 70 or 80 Leagues, you allow not the Variation, but ſteer Eaſt or Weſt by Compaſs, you ſhall fall to the Northwards or Southwards of your Port on each 19 Leagues of Diſtance, one Mile for each Degree of Variation, which may produce very dangerous Errors, where the Variation is conſiderable; for Inſtance, having a good Obſervation in Latitude 45d. 40m. about 80 Leag. without *Scilly*, and not conſidering that there is ⅛ Degrees Weſt Variation, I ſteer away *Eaſt* by Compaſs for the Channel; but making my way truly E. 8d. *N.* when I come up with *Scilly*, inſtead of being 3 or 4 Leagues to the South thereof, I ſhall find my ſelf as much to the Northward: And this Evil will be more or leſs according to the Diſtance you fail in the Parallel. The Rule to apply it is, That to keep your Parallel truly, you go ſo many Degrees to the Southward of the *Eaſt*, and Northward of the *Weſt*, as in the *Weſt* Variation; but contrariwiſe, ſo many Degrees to the Northwards of the *Eaſt*, and Southwards of the *Weſt*, as there is Eaſt *Variation.*

A further Uſe is in many Caſes to eſtimate the Longitude at Sea thereby; for where the *Curves* run nearly *North* and *South*, and are thick together, as about Cape *Bona Eſperance*, it gives a very good Indication of the Diſtance of the Land to Ships come from far; for there the Variation alters a Degree to each two Degrees of Longitude nearly; as may be ſeen in the *Chart.* But in this Weſtern Ocean, between *Europe* and the *North America*, the Curves lying nearly Eaſt and Weſt, cannot be Serviceable for this Purpoſe.

This Chart, as I ſaid, was made by Obſervation of the Year 1700, but it muſt be noted, that there is a perpetual tho' ſlow Change in the *Variation* almoſt every where, which will make it neceſſary in time to alter the whole Syſtem: at preſent it may ſuffice to advertiſe that about *C. Bona Eſperance*, the Weſt Variation encreaſes at the Rate of about a Degree in 9 Years. In our Channel it encreaſes a Degree in ſeven Years, but ſlower the nearer the Equinoctial Line; as on the *Guinea* Coaſt a Degree in 11 or 12 Years. On the *Americas ſide* the *Weſt Variation* alters but little; and the *Eaſt Variation* on the *Southern America* decreaſes, the more Southerly the ſaſter; the *Line of No Variation* moving gradually towards it.

I ſhall need to ſay no more about it, but let it commend it ſelf, and all knowing Mariners are deſired to lend their Aſſiſtance and Informations, towards the perfecting of this uſeful Work. And if by undoubted Obſervations it be found in any Part defective, the Notes of it will be received with all grateful Acknowledgement, and the Chart corrected accordingly.

E. HALLEY.

This CHART is to be fold by *William Mount*, and *Thomas Page* on *Tower-Hill.*

*Figure 6.3.* Isogonic map of the Atlantic by Edmond Halley (1701).

means of cartographic representation in other connections later in this work (appendix B).[13] In 1702, Halley published a larger map that extended the isogones to the Indian Ocean (based on the observations of others, including William Dampier and East India Company navigators) but not to the Pacific, for which no adequate data were yet available. Both Halley's Atlantic and world isogonic charts feature rococo-style cartouches but generally have less decoration than is usual for this period.

A third map by Halley links the thematic with the *hydrographic* tradition; it again illustrates the interest of this scientist (whom his contemporaries considered second only to Newton among the English natural philosophers) in cartographic work. Before discussing this map in particular, we should briefly review the progress of marine cartography up to this point. The delineation of coastlines (separating those two fundamental geographical quantities, land and water) had, of course, been part of cartography since antiquity, and in the portolan chart of the later Middle Ages, we have seen a cartographic device specifically designed for the navigator. We know that this type of chart formed the basis of early Iberian mapping, but, in time, as the European view of the world expanded through geographical discoveries, the character of the sea chart also underwent change. To assist seamen, various navigational manuals were published, some of which contained views of coastal features and, later, charts. We have discussed the contributions of sixteenth-century northern Europeans (especially the German and Low Country cartographers) to mapping, the high point of which was the development of a chart of particular use to the navigator—the Mercator—in 1569.

Nearly sixty years after this, Johannes Kepler (1571–1630), to whom Newton and all later astronomers were indebted for the discovery of the elliptical paths of planets (rather than perfectly circular orbits, as previously believed), published his Rudolphine Tables of planetary and star positions based on the observations of Tycho Brahe (1546–1601). By observing the rim of the moon in relation to a known star, or a lunar eclipse, it is possible to calculate longitude at the point of observation by calculating local time and comparing it to the time stated in Kepler's tables. On this basis, a world map was made by Philipp Eckebrecht and published in Nürnberg in 1630, with its prime meridian at Tycho Brahe's observatory on Hveen Island, Denmark. This is the first map to establish longitude by difference in time (one hour equals 15 degrees longitude) through observation of celestial phenomena. With the greatly improved longitudinal position provided by such maps, further progress could be made.

Halley, who combined both practical and theoretical qualities, began marine surveying a decade before he took command of the *Paramore*.[14] He made a chart of the mouth of the River Thames and, five years later, one of the Sussex coast. Upon returning from his Atlantic voyages, Halley secured permission to use the *Paramore* for the purpose of surveying the English Channel, and in 1702 he published a map resulting from this activity (fig. 6.4). Superficially, this map resembles other charts of the period, with its representation of coastlines (understandably much improved over those of the same area in Ortelius's or Waghenaer's work of over a century earlier), shoals, anchorages, depths (marked in fathoms), and so on. Even the compass roses with radiating lines are reminiscent of earlier marine charts, but Halley's map differs from these in at least two particulars. First, Halley provides a formula for estimating the height of the tides at certain places, which are indicated on the map by Roman numerals (the direction of tides being shown by arrows). Second, Halley originated the resection method of coastal survey, which can be accomplished on a ship under sail, and for greater accuracy took angles by the sun rather than by magnetic compass, as was the usual practice at that time. Halley gave up his active naval command at about the time of the publication of his tidal chart but continued to interest himself in cartography to the end of his life. In 1715 he produced a map of the shadow of the moon over England resulting from the total eclipse of the sun; the time taken by the passage of the moon's shadow is also indicated. This map beautifully illustrates the concern of Halley—who became Astronomer Royal at Greenwich Observatory in 1720 in succession to John Flamsteed (1646–1739)—with the entire cosmos. Since it was made before the event it depicts, this map demonstrates the highest attribute of science: the ability to predict. Like Halley's other major thematic maps, it is a work of great originality.[15]

We have discussed the isogonic maps of Halley and indicated that his Atlantic chart of 1701 is presumably the earliest published isoline map. However, two manuscript maps with isobaths (lines of equal depth of water), about which he could not have known, antedate Halley's map. They are by Pieter Bruinss (1584) and Pierre Ancelin (1697). Apparently no cartographer between these dates thought of joining up points of equal depth. A scattering of depth values along coasts, a feature of the maps of the Renaissance cartographers, was made possible by that age-old navigational instrument, the lead and line for plumbing depths. However, in 1729, an engraved map of isobaths was published (fig. 6.5). It was the work of Nicholas Samuel Cruquius (Cruquires) (1678–1754), a Dutch engineer, and shows depths in the Merwede River, a distributary

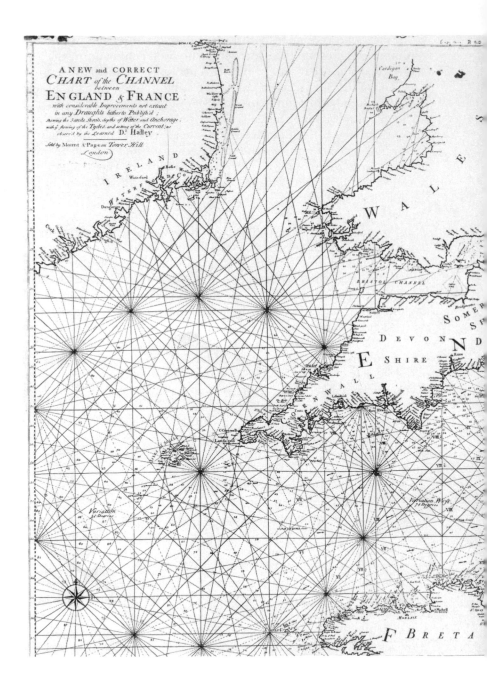

A NEW and CORRECT
CHART of the CHANNEL
between
ENGLAND & FRANCE
with considerable Improvements not extant
in any Draughts hitherto Publish'd ;
shewing the Sands, Shoals, depths of Water and Anchorage ;
with y. flowing of the Tydes, and setting of the Current ; as
observ'd by the Learned Dr. Halley

Sold by Mount & Page on Tower Hill
London

*Figure 6.4.* Chart of the tides in the Channel between England and France by Edmond Halley (1702).

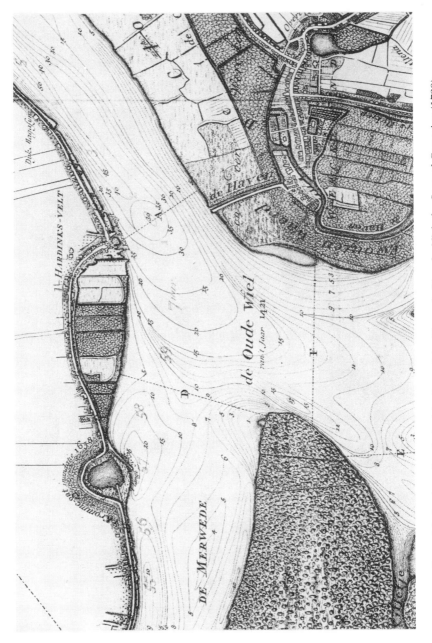

*Figure 6.5.* Section of an isobathic chart of the Merwede River by Nicholas Samuel Cruquius (1729).

of the Rhine.[16] Depth values are marked in the main river, but in the smaller tributaries and other water bodies, the usual unquantified form lines of the period are drawn (compare fig. 6.5 with fig. 6.9); in the latter case, the water lining off the coast is nonquantitative.

Prior to this, with the approval of the Académie Royale des Sciences and using fixed points provided by the astronomers of the Paris Observatory, Jean Picard (1620–82) and Gabrielle Phillipe de la Hire (1640–1718) surveyed the coasts of France. Detailed charts arising from the new French marine survey, drawn on the Mercator projection with latitude and longitude indications, were published later as *Le Neptune français, ou atlas nouveau des cartes marine* (1693). Great discrepancies were found between the coast or shorelines established by this marine survey and their representation on the best previous maps of France, those of the Sanson family, as shown in figure 6.6.[17] The French were the first to eliminate the redundant compass roses from their charts, and it was they who established a single official hydrographic office, the Dépôt des Cartes et Plans de la Marine, in 1720. Some time after this, in 1737, an isobathic chart was made of the Channel *(La Manche)* by Philippe Buache (1700–73).[18] At a time when official French hydrographic surveying was progressing, the English were still relying on private or quasi-official charting throughout the seventeenth and most of the eighteenth centuries. Thus members of the Thames School—William Hack (flourished 1650–1701), John Seller (fl. 1669–97), John Thornton (fl. 1650–1700), and so forth—made maps and atlases of the coasts of the world, while Greenvile Collins (fl. 1669–98) surveyed the British shorelines and produced engraved charts of the area as well as *Great Britain's Coasting Pilot* (1693), the first such atlas published in England.[19] An interesting manuscript atlas of this period is *The South Sea Waggoner,* consisting of over one hundred charts of the Pacific coasts of the Americas captured from the Spanish in a sea battle by English pirates; when they returned to England, one of them, Basil Ringrose (ca. 1653–86), edited the collection.[20] The East India Company, founded in 1600, maintained its own staff of hydrographers, of which the best-known is Alexander Dalrymple (1737–1808). These activities were buttressed by the publication of marine books by the firm of Mount and Page, which had its origins with William Fisher (1622–92) and continued until the end of the eighteenth century.[21]

In the second half of the eighteenth century there was an increased interest in charting the coasts of the world, particularly of those areas only recently discovered by Europeans. Outstanding charts of New Zealand, part of Australia, and North America, as well as of Tahiti, Hawaii, and many other Pacific islands, were produced by Captain James

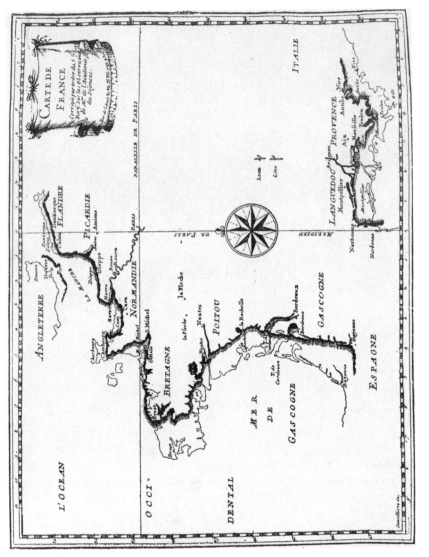

*Figure 6.6.* The rendering of the coastlines of France resulting from surveys of the scientists of the Académie Royale in 1693 (shaded), superimposed over the delineation of Sanson from 1679 (line).

Cook.[22] Cook had his training as a marine surveyor in the Saint Lawrence River region of eastern Canada, and surveys of this area produced in 1758–59 helped General James Wolfe in his defeat of the French at the Battle of Quebec. But it was on his first Pacific voyage (1768–71) that Cook had greater opportunities to display his talents in producing running surveys of coasts not previously charted. The first object of Cook's voyage was to observe the transit of Venus, whose future occurrences were given in a table published by Halley in 1716.[23] While this observation was being successfully accomplished, Cook made "A Plan of King Georges Island or Otaheite" in 1769 (fig. 6.7) and assisted in another showing the whole of the Society Islands, which was mainly the work of Tupaia, the Polynesian navigator whom Cook employed. When Cook reached New Zealand, he made coastal surveys of those islands that had been visited in 1644 by the Dutchman Abel Tasman, who partially mapped New Zealand's west coast. Even more important than Cook's survey of all the coasts of these islands was his charting of the east coast of Australia (1770), hitherto unmapped by Europeans, which Cook called New South Wales. On a second Pacific voyage between 1772 and 1775, Cook circumnavigated the globe in the midlatitudes of the Southern Hemisphere and proved that there was no great continent in that area, as had been postulated since antiquity. As on Cook's previous Pacific voyage, much useful charting resulted, helped, in this case, by the marine chronometers aboard the ships. These were copies of one by John Harrison (1693–1776), who made timepieces of sufficient accuracy to "keep" longitude at sea, for which he won the great prize of twenty thousand pounds from the British government.[24] A third Pacific voyage proved fatal to Cook, who in 1778 discovered and sojourned in Hawaii. Apparently these islands had been missed by the Spanish galleons that had plied annually across the Pacific from Acapulco, Mexico, to Manila in the Philippines and back after Andrés de Urdaneta pioneered these voyages in 1565. The purpose of Cook's projected third circumnavigation was to seek the elusive Northwest Passage from the Pacific. After reconnaissance mapping part of the northwest coast of North America and failing to find the passage, Cook returned to Hawaii, where he was killed in 1779. Work on surveying the northwest American coast was continued between 1792 and 1795 by George Vancouver, who had accompanied Cook on his second and third voyages of discovery, and surveys of the coasts of Australia were made between 1798 and 1803 by Matthew Flinders (177?–1814).[25]

Though contemplated earlier, it was not until 1795 that an official British Hydrographic Office was founded, of which Dalrymple, who had been denied the command of what became Cook's voyages, was appointed first hydrographer.[26] Meanwhile, the French under Louis-Antoine de Bougainville (1729–1811) and Jean-François de Galaup, Comte de la Pérouse (1741–ca. 1788), and others added significantly to the charting of the Pacific. The Spanish mounted scientific charting expeditions in the eighteenth century, such as that of Alejandro Malaspina (1754–1810) from 1789 to 1794, as earlier Peter the Great and his successors had with their sponsorship of the Dane Vitus Bering's journeys (1728–41). Increasingly, as the eighteenth century progressed, Russians entered the field of charting the coasts and mapping the land in their vast domain.[27]

It is understandable that isobaths, which show the configuration of a surface (in the case of Cruquius's map a riverbed, a concept rather different from isogones), should have been the first form of contour lines applied to delineate any part of the lithosphere. The surface of the water, though variable in height through time within certain limits, forms a convenient and natural datum (base to which measurement can be referred), and the lead and line provides a relatively easy means of gathering the needed depth information in shallow areas. It was several decades before the contour principle was applied in a significant way to dry land surfaces, partly because of the difficulty of making the necessary measurements and partly because of the preference on the part of cartographers and map users for the hachure technique of landform representation, which will be discussed subsequently.[28]

During the seventeenth and eighteenth centuries, France became the leader in *topographic mapping*, developing methods that became standard and were later widely adopted elsewhere. This began after the astronomer Giovanni Domenico (later Jean-Dominique) Cassini (1625–1712), who was a professor at Bologna, accepted an invitation to the Académie Royale in Paris. As mentioned earlier, both that society and the Royal Society of London were founded in the mid–seventeenth century, and both concerned themselves with a wide variety of scientific problems, including mapping and charting.

Long before this time the mapping of countries and smaller land areas was of course undertaken. We have illustrated this with Humphrey Lhuyd's map of England and Wales from Ortelius's *Theatrum* (see fig. 5.10 in chap. 5). More detailed maps of this area—the county maps of Christopher Saxton and John Norden, and the estate maps, or terriers, by a number of cartographers—had also appeared. Examples could be provided from a number of European countries where progressively larger-scale and more detailed maps were made, so that by the end of the sixteenth century regional maps, some consisting of multiple sheets, had existed for a good many years. In some instances, such as the work

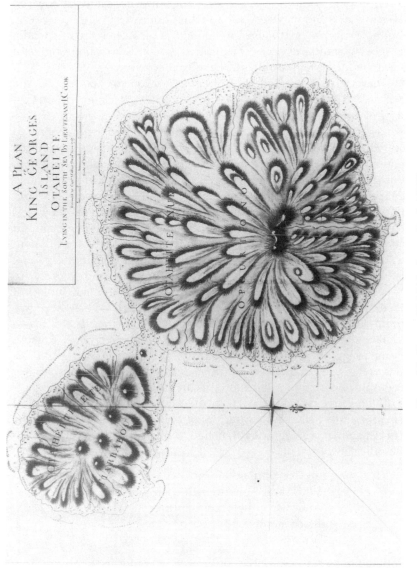

*Figure 6.7.* Chart of Tahiti by James Cook (1769).

of Willebrord Snell (Snellius) (1580–1626), who utilized the method of determining distances by trigonometric triangulations, real progress was made in accuracy through better techniques and improved instruments. However, if maps with even larger and still uniform scale that covered an extensive area and that were suitable for administrative, engineering, and military purposes (topographic quadrangles) were to be made, more rigorous standards had to be applied. Cassini's arrival in France in 1669 initiated the topographic survey of that country.[29]

We have already alluded to the *planisphère terrestre* laid out on the floor of the Paris Observatory by Cassini; interestingly, Halley visited Cassini at the observatory in 1682, when he was engaged in this work. Cassini's master map of the world (with an azimuthal projection centered on the North Pole) recalls the *Padrón General* of the Casa in Spain or Ptolemy's maps in that an attempt was made to collate all current geographical data. But unlike these works, in which the location of places depended to a large extent on verbal information from travelers by land or upon dead-reckoning estimates supplied by sailors, no place was added to Cassini's planisphere unless its position had been determined astronomically. This information was published in map form by Cassini in 1696, but it also became well known through the compilations of Guillaume de Lisle (1675–1726) and other cartographers who had access to the master map in the Paris Observatory. In this way, the true length of the Mediterranean—approximately 42 degrees, which we have seen was accurately determined by Arab astronomers by the twelfth century—was, as far as we know, first correctly recorded on printed maps. Cassini, following a suggestion of Galileo, also prepared tables of the revolutions of the satellites of Jupiter, so they could be used for the determination of geographical longitude. Through this method, knowledge of the locations of many areas was improved. At the same time, much conjectural information, especially on the interiors of continents, was eliminated. It was this imaginary cartography that had evoked Jonathan Swift's well-known satirical comment:

> So Geographers, in Afric-maps,
> With savage-pictures fill their gaps;
> And o'er unhabitable downs
> Place elephants for want of towns.
>
> (*On Poetry,* line i.117)

A detailed and accurate map of France in multiple sheets and employing uniform standards and symbols was needed; at the request of Jean-Baptiste Colbert (1619–83), Louise XIV's finance minister, and with royal support, the Académie under Cassini attempted to meet this challenge. The first step was to measure the arc of the meridian of Paris to ascertain the length of a degree of latitude. This was undertaken by the Abbé Picard by means of triangulation, a method of which he was a strong advocate. Picard used a quadrant with telescopic sight and filar micrometer of his own invention. The work, completed in 1670, was the basis for a series of nine topographic sheets of the Paris area made by David du Vivier. Picard died in 1682, but subsequently the meridian of Paris was extended from the Channel to the Pyrenees. A practical result of this was that, with extension, the triangulation could be the basis of more accurate topographic maps of an entire country than had ever been produced before. In the more theoretical realm, measurement of the degrees of latitude over a long north–south line cast doubt, ironically, upon Picard's belief in a perfectly spherical earth and on the prolate spheroid theory supported by Cassini's son, Jacques, and others. However, further measurements sponsored by the Académie near Quito—directed by Louis Godin (1704–60), Pierre Bouguer (1698–1758), and Charles Marie de la Condamine (1701–74)—and measurements in Lapland by Alexis-Claude Clairaut (1713–65), Pierre-Louis Moreau de Maupertuis (1701–74), and others confirmed the general correctness of Newton's hypothesis of the earth as an oblate spheroid. Interestingly, it was a comparison of observations on the length of the pendulum in equatorial South America by the Frenchman Jean Richer (1630–96) with measurements in Paris and elsewhere that formed the basis of Newton's postulation.

After the death of his father, Jacques Cassini de Thury (1677–1756) continued the survey of France and succeeded his father as the head of the Paris Observatory. The meridian was resurveyed with refined methods and triangulation extended east and west. Figure 6.8 shows a small section of the map of the triangulation of France by Giovanni Domenico (later Jean-Philippe) Maraldi (1709–88) and Jacques Cassini in 1744, the whole work consisting of some forty thousand triangles. Cassini was assisted by his son, César-François (1714–84), who carried the topographical mapping of France, including the filling in of detail, to a virtual conclusion by subscription alone, without official financial support. The few sheets not completed before his death in 1784 were prepared under the direction of his son, Jean (1748–1845). The topographical map of France—182 sheets on the scale of "*une ligne pour cent toises*" (1:86,400)—was eventually completed in 1793. Thus, four generations of the Cassini family over a period of more than one hundred years had been involved in the first true topographic survey of an entire country, in which the

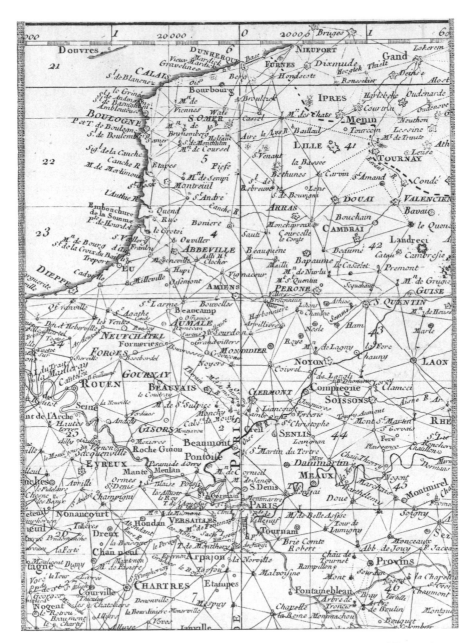

*Figure 6.8.* Section of the map of the triangulation of France by Giovanni Domenico (Jean-Philippe) Maraldi and Jacques Cassini de Thury (1744).

principle of providing a rigorous framework for the whole survey before the details were filled in was applied. The sheets, when all assembled, form approximately a thirty-six foot square.

Figure 6.9 is a small sample of one of these well-engraved topographical maps, showing part of the northeast coast of France. The symbolization is essentially the same as that employed on the Paris sheets of over a century before. In particular, the terrain representation, which indicates two or at most three surface levels, is unsatisfactory. Hachures (short lines whose thickness indicates steepness of the slope) were used to delineate these tabular surfaces. Hachuring was systematized in 1799 by the Saxon topographic engineer Johann G. Lehmann (1765–1811), but it remained essentially a qualitative method of landform representation in the sense that absolute elevation cannot be read from hachures alone. Hachures are drawn downslope rather than around the features, as are contours. A three-dimensional effect, simulating illumination from the northwest, was produced by emphasizing the hachures on the south and east slopes on the maps of Switzerland by Guillaume-Henri Dufour (1787–1875). At best the technique is expressive, but at worst the hachures degenerate into "hairy caterpillars." In addition to the hachures on the *Carte de Cassini,* some obliquely viewed landforms appear, such as the coastal sand dunes in figure 6.9. These, and the symbols for smaller settlements and forests, are not in plan view, in contrast to the representation of larger settlements and, of course, roads, rivers, coastlines, and so on.

Delineation of the continuous three-dimensional form of the land has always been one of the most challenging problems in cartography. As we have seen, from earliest times, so-called "fish scales," "molehills," or "sugar loaves" in profile or, at best, oblique views have been used to represent relief. Such forms utilize planimetric position in two dimensions. The planimetric displacement that results can apparently be tolerated on small-scale maps but not on those of large or topographic scale. Features hidden by the landforms can, of course, be realigned only by destroying the accuracy of the map. What was needed was a planimetrically correct quantitative method of rendering terrain, and such had actually been developed before this time in isometric lines of the sort used by Halley or Cruquius.

The measurement of elevation of land features had been possible ever since the development of trigonometry and the invention of instruments such as the altazimuth theodolite. However, it was laborious work to survey the large number of points needed for an accurate isometric (contour) map of the land surface with this instrument. For a time it seemed that the barometer, developed by Evangelista Torricelli (1608–47) in 1643, might provide an easy means of producing the needed data, as the lead and line had for isobathic maps of river basins. Blaise Pascal (1623–62) demonstrated that atmospheric pressure decreases as a function of increasing elevation by having a barometer taken up the Puy-de-Dôme in the Massif Central of France, and he also made the famous statement that we live at the bottom of a sea of air. This principle was employed in cartography by an English physician, Christopher Packe (1686–1749). In the compilation of his Philosophico-Chorographical chart of Kent, published in 1743, Packe used barometric readings converted to elevation figures and plotted these as spot heights from a datum provided by high tide in the Channel.[30] However, in spite of his precaution of taking readings only on days with similar weather, the temporal variation of pressure rendered this method less than satisfactory. On the *Carte de Cassini,* some spot heights appear. A number of modest attempts to draw contours (lines connecting all points with the same altitude) were made by military engineers in the second half of the eighteenth century. An interesting early attempt at contouring appears in a map of the terrain of Oxford, where from the highest point in the city (Carfax) the contours are referred downward, in the manner of isobaths, rather than upward from a datum such as sea level. The earliest use of contours covering a large area on a topographic map is attributed to a French engineer, Jean Louis Dupain-Triel (1722–1805).[31] The date of this work is 1791, although he later added tints between selected contour levels to produce a colored hypsometric (altitude-tint) map. But it was not until the middle of the nineteenth century that contouring displaced the hachure method of relief representation in many of the great national surveys. In some cases contours were drawn as a basis for hachuring and then removed from the final sheets after the hachures were drawn.

The value of detailed topographic mapping of countries was quickly appreciated by administrators. Very soon after its publication, the *Carte de Cassini* was taken over by the French government. Later, in the hands of Napoleon Bonaparte, who had a great appreciation of geography, mapping because a prime instrument of administration and conquest. Meanwhile, in Britain, General William Roy (1726–90), who had long advocated the detailed mapping of that country, embarked on a trigonometric survey in 1783. It is from this beginning that the official, quasi-military Ordnance Survey was formed.[32] The British and French triangulation networks were connected across the Channel in 1787. A quarter of a century before this, the French survey and the triangulation of the Netherlands by Snell (noted earlier) had been joined overland.

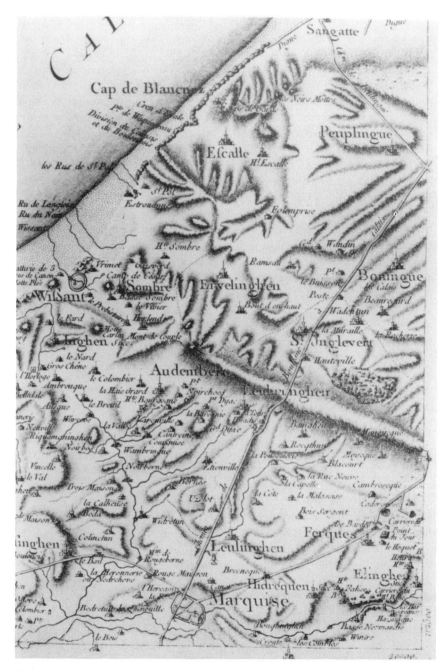

*Figure 6.9.* Section of engraved topographic map of France from the Cassini surveys.

Some of the most interesting cartographic developments were soon to take place in areas where there were fewer restraints, away from the main population centers of northwestern Europe. In Ireland, for example, instead of using one meridian, as had been done in France, the Ordnance Survey employed several meridians to further reduce error in mapping. Leveling, also introduced in the surveys of this area, permitted contours to be drawn with greater accuracy and facility. In India, surveying on the European model was begun in 1765 in Bengal by James Rennell (1742–1830). Triangulation was initiated in Madras in 1802, which led to the Great Trigonometrical Survey of India, begun by William Lambton and completed in 1843 by Sir George Everest (1790–1866), which takes us beyond the time constraint of this chapter. However, not without great travail, the Survey of India eventually produced topographic maps of essentially all of the subcontinent, so that it could later be claimed that India was the best-surveyed large country in the world.[33] Much was also eventually learned about the shape of the earth and about earth magnetism through these surveying activities.

One of the results of the French Revolution was the reform of weights and measures. In 1791 the Paris Académie des Sciences, formerly the Académie Royale des Sciences, defined the meter as 1/10,000,000 of the quadrant of the terrestrial meridian, and various countries converted their standard linear measure to the metric system. This led to the concept of the "natural scale" in cartography, whereby one unit of length on the map is represented by a given number of like units on the earth. This so-called representative fraction (R.F.) was first used in France in 1806. The representative fraction—for example, 1:63,360—is a more general expression of scale than its verbal equivalent, "one inch equals one mile." These two scale expressions plus the graphical scale are in common use today, as well as, occasionally, an area scale.

In the North American colonies there had been a great concern with mapping of the land since the earliest European settlements.[34] In 1585, John White (fl. 1585–93) drew a map of the Cape Hatteras area during the second Roanoke expedition sponsored by Sir Walter Raleigh. White, who painted excellent watercolor renderings of the "Indians" and their settlements in what was then part of Virginia, was most probably aided in his mapping endeavors by Thomas Hariot or Harriot (1566–1621), a brilliant mathematician and naturalist. The map was published in Hariot's *A Brief and True Report of the New Found Land of Virginia,* in part one of the *Grand Voyages,* engraved by Theodore de Bry (1528–98), which provided a window on America for many Europeans. A contemporary of White and Hariot was the Flemish cosmographer André Thevet (1502–90), who drew maps of parts of America based on his own travels.[35]

Some years later (1608–9) Captain John Smith explored Chesapeake Bay and produced an extremely influential map of that locality, which went through many printings in the seventeenth century. On Smith's map the help of the indigenous population is acknowledged by a series of crosses that separate those areas he had visited from those he had "by relation" with the Indians, an early example of "reliability" on a map. One of the most important American maps of colonial times was that produced in 1670 by Augustine Hermann (a Bohemian) for Lord Calvert of his Maryland patents and a large region beyond.[36] A map covering an even larger area from Virginia to the Great Lakes was made by Joshua Fry (under whom George Washington served in the French and Indian Wars) and Peter Jefferson (father of Thomas Jefferson) in 1751.[37] Both George Washington and Thomas Jefferson, like many of their contemporaries in America, engaged in surveying and had a great interest in mapping and geographical exploration. We have concentrated on Virginia and the middle colonies, but there was an equally high interest in mapping in other areas, including the northern colonies, from an early date. John Smith drew a map of New England that was published in 1614, to be followed by other maps of this area by William Wood (1635) and by John Foster (1677). The latter, a woodcut, has the distinction of being the first map printed in Anglo-America. The French were also active with "Novelle France" by Marc Lescarbot (1609), the earliest map to show Quebec, and Samuel de Champlain's map of the same area (1632) based on his explorations of the lower Great Lakes. More of these extensive inland water bodies is shown on maps in the *Jesuit Relations* (1670–71) and in a map of New France by Louis Joliet (1645–1700) of the explorations he and Jacques Marquette (1637–75) undertook between 1672 and 1697.

The most ambitious map made of colonial America was that by John Mitchell, dated 1755 (fig. 6.10). Mitchell (d. 1768) was a physician who was born in Virginia and received his medical education in Edinburgh. He returned to Virginia but emigrated in 1746 to England, where he was commissioned to make a map to define British territorial rights in North America. Twenty-one editions or impressions of Mitchell's map appeared in four languages between its initial appearance in 1755 and 1781. Shown on the map are the boundaries of colonies, a number of which remain as the limits of states of the United States today. Some were resurveyed, such as the line between Pennsylvania and Maryland

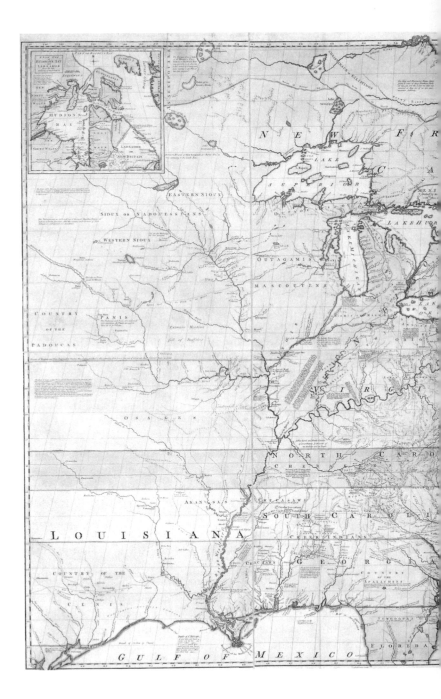

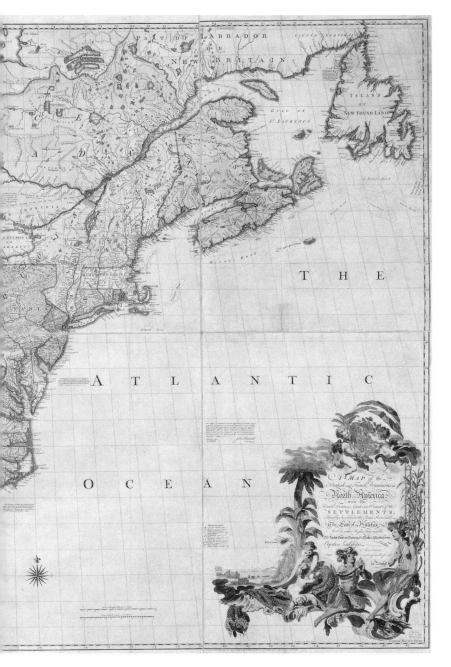

*Figure 6.10.* John Mitchell's map of British and French dominions in North America, with western claims of the colonies indicated (1753).

by the English astronomers Charles Mason and Jeremiah Dixon in 1768, which nearly a century later became "Dixie," the boundary between the South and the North in the American Civil War. Other boundaries were eliminated, including those showing claims to western lands that were expunged by the establishment of the public domain (1784) and the creation of new states. A copy of Mitchell's map showing the claims between the United States and what was to become Canada was used in the negotiations for the 1783 Treaty of Paris, which well illustrates the use of the map as an important political instrument.[38] Although during this period most maps of America were printed in Europe, there were, after the middle of the seventeenth century, some notable exceptions, including Lewis Evans's map "The Middle British Colonies, 1755." This was engraved by James Turner and is believed to have been printed by the press of Benjamin Franklin.

Franklin, a fellow of the Royal Society, with his cousin Timothy Folger, a sea captain, produced a chart in 1775 showing for the first time the limits of the Gulf Stream, based on temperature from Fahrenheit thermometer readings.[39] Another outstanding hydrographic contribution is a chart of Boston Harbor from the *Atlantic Neptune* (1777–81) by Joseph Frederick Wallet des Barres (1722–1824). This atlas contains 182 charts and views. In spite of the attention to the cartography of the area over the preceding 150 years, good maps were not available to the colonists in the Revolutionary War, as recognized by George Washington, who took steps to remedy this situation. Even after the war some American maps continued to be published in England, including Thomas Jefferson's "Map of the Country between Albemarle Sound and Lake Erie, 1787."[40] But the seeds of a distinctive American school of cartography, in which, as in Europe, the military surveyor played a major role, were sown during the war.[41]

A number of mathematicians in the seventeenth and eighteenth centuries turned their attention to the study of map projections. They built upon an inheritance that included such models as the orthographic, stereographic, plane chart, Ptolemaic (three projections), cordiform (many cases), and Mercator (three projections), as detailed earlier. The sinusoidal projection was in fairly regular use through the earlier part of the seventeenth century and later. In the publications of the Sansons, a French family of cartographers (active 1650–1700), this projection was so widely used that their name became attached to it, as did that of Flamsteed, who employed it for his star charts. Today it is frequently called the Sanson–Flamsteed projection (fig. 6.11a). C.-F. Cassini invented a projection (fig. 6.11c) that was used for the French topographic surveys until 1803, when it was replaced by the Bonne (fig.

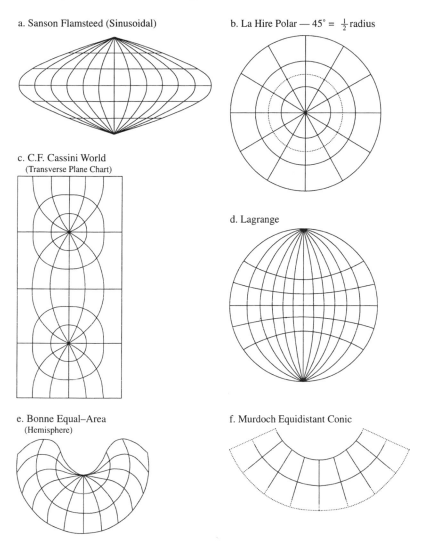

a. Sanson Flamsteed (Sinusoidal)

b. La Hire Polar — 45° = ½ radius

c. C.F. Cassini World
(Transverse Plane Chart)

d. Lagrange

e. Bonne Equal–Area
(Hemisphere)

f. Murdoch Equidistant Conic

*Figure 6.11.* Projections devised in the seventeenth and eighteenth centuries, shown with a 30-degree graticule of latitude and longitude.

6.11e), popularized by Rigobert Bonne (1727–95), who had in fact revived a projection with antecedents in the early sixteenth century going back as far as Peter Apian. Bonne's projection is based on a central meridian that is true to scale, with all of the parallels also being true to scale as well as concentric circles. Other interesting projections invented or adapted in the eighteenth century include the perspective projection of

Transverse Cylindrical Equal-Area

Polar Azimuthal Equal-Area

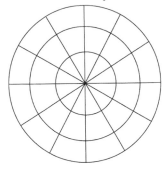

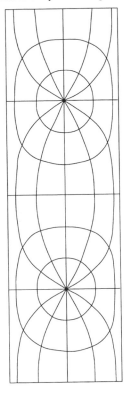

Conic Conformal With Two Standard Parallels

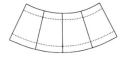

Equatorial Azimuthal Equal-Area

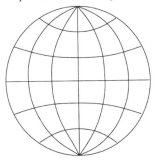

Conic Equal-Area

Transverse Mercator

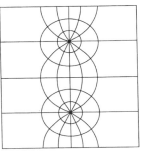

Cylindrical Equal-Area

*Figure 6.12.* Projections of Johann Heinrich Lambert (1728–1777), shown with a 30-degree graticule of latitude and longitude.

Philippe de la Hire (1640–1718), the globular projection of Joseph Louis Lagrange (1736–1813), and the equidistant conic projection of Patrick Murdoch (d. 1774), a clergyman and not the only man of the cloth to devise useful projections (figs. 6.11b, d, and f).[42]

Undoubtedly the most prolific inventor of projections in the eighteenth century was Johann H. Lambert (1728–77), even though his general contributions to science were overshadowed by his fellow Swiss-born mathematician Leonhard Euler (1707–83), who was also interested in the transformation of the all-side-curving figure of the globe on a plane surface.[43] Lambert invented a number of projections that are in use today (fig. 6.12). His output included the transverse cylindrical equal-area, the conic equal-area, the cylindrical equal-area, the polar azimuthal equal-area, the conic conformal with two standard parallels, the equatorial azimuthal equal-area, and the transverse Mercator. This last projection, often known as the Gauss conformal (and thus wrongly attributed to the famous German mathematician Carl F. Gauss, 1777–1855) has recently been put to important uses and will be discussed subsequently. It will be noticed by these names that most of Lambert's projections are "equal-area" (equivalent), a property that is mutually exclusive with "conformality" (correct shape around a point). Equivalence was a quality that would be in increasing demand in thematic, statistical, distributional maps in the nineteenth and twentieth centuries, and thus several of Lambert's projections were available for this purpose before they were needed.

Globe making (terrestrial and celestial) continued during this period in the Netherlands by the Blaeus and others, but eventually these workers were surpassed by the Venetian Vincenzo Coronelli (1650–1718). Coronelli, a Franciscan friar who founded the first geographical society, the Academia Cosmographica degli Argonauti, in 1684, contributed to many branches of cartography. However, he is best known for his globes, large and small, and for *Libro dei globi,* a printed atlas of globe gores (1701).[44] In England in the 1670s, Joseph Moxon (1627–1700) was making terrestrial pocket globes three inches in diameter, each encased in a celestial globe. This led to the invention of "globe machines" to display planetary motions called *orrery* after Charles Boyle, Earl of Orrery (1676–1731), for whom such an instrument was made.

As we have seen, some of the greatest figures in Europe and America in the seventeenth and eighteenth centuries were connected with cartographic representation: Galileo, Kepler, Cassini, Petty, Newton, Halley, Euler, Cook, Franklin, Washington, and Jefferson. During this period before the age of specialization, however, some such figures may have spent only a small fraction of their time in this endeavor. It was the "lesser men," as Eva Taylor called them, as well as women who took care of the day-to-day business of the map trade and occasionally made a breakthrough.[45] As we turn to the nineteenth century, we will find a greater degree of specialization, but there was still room for the polymath and the gifted amateur as well as the professional in the field of cartography.

SEVEN

# Diversification and Development
# in the Nineteenth Century

Some immediate effects of the Scientific Revolution upon cartography were discussed in the previous chapter. The cartographical development of other important concepts propounded in the seventeenth and eighteenth centuries took place during the Industrial (and Technological) Revolution that followed. Thus the foundations of modern statistical methods were laid in the seventeenth century by such workers as Petty, who investigated demographic problems and popularized the study of vital statistics, while his contemporaries Huygens and Halley concerned themselves with the theory of probability that led, in the eighteenth century, to the use of accurate statistical methods.[1] The holding of regular censuses—which began (in modern terms) in Sweden in 1749, in the United States in 1790, and in Britain in 1801—provided a large potential source of mappable data. There was also a great increase in knowledge of the physical world at this time. The net result was that the first half of the nineteenth century, in particular, was a period of rapid progress along a broad front in mapping, especially in thematic mapping. Commercial and government map publishing increased greatly, leading to the expansion of existing facilities and the creation of new ones. Atlases, notably the *Hand Atlas* (1817–22, in fifty sheets) by Adolf Stieler (1775–1836), and wall maps such as those of Emil von Sydow (1812–73) made the new material available to large numbers of students and to the general public. The uses of globes, maps, and atlases also became important school subjects for both girls and boys at this time.

We have discussed the beginnings of modern topographic mapping in France and its extension to other countries, notably Britain and its overseas empire, especially India. Before the 1860s, some states of the then-divided greater Germany and Italy, looking to France for leadership in many aspects of culture, followed suit. During the nineteenth century, other countries in Europe and beyond established official topographic surveys, often as part of the army's responsibilities or, at least, with military ends in view. But topographic maps, which are a fundamental primary source of geographical information, also provide a base for other data most meaningful when expressed at large scale—for example, land use and geology. Broad categories of land use had been shown on early estate and county maps by various hand-painted and engraved symbols, and Christopher Packe had distinguished between arable land, downland, and marsh in his map of 1743, which was discussed previously. A more ambitious and systematic attempt in this cartographic genre was Thomas Milne's land-use map of London in 1800, drawn at the detailed scale of two inches to the mile. Milne used colors (applied by hand) and letters to indicate some seventeen categories of land use, producing a map not unlike those of the Land Utilisation Survey of Britain of over a century later.[2] Land-use maps show, directly or indirectly, the surface cover—crops, forests, urban forms, and so on—at a given time. These data are ephemeral and in this respect contrast with the more permanent information on geological maps, which show rock formations with the overburden stripped away.

Although tailing piles give archaeological evidence of an interest in minerals before the written record appears, geology in the modern sense was founded at the end of the eighteenth and the beginning of the nineteenth century through the efforts of a number of scientists, including James Hutton (1726–97) in Scotland, Abraham Werner (1749–1817) in Germany, and the Baron Georges Leópold Chrétien Dagobert Cuvier (1769–1832) in France. But it is a contemporary of these workers, the English civil engineer William Smith (1769–1839), who is credited with first adequately correlating fossils with associated strata and then developing geological mapping.[3] Smith's cartographic efforts were more ambitious and more successful than any previous attempts in this direction. In 1815, after nearly a quarter of a century of research and observation, he published his map "The Strata of England," a small part of which is shown as figure 7.1. This hand-colored map is all the more remarkable when we realize that the data were gathered by noting the strata along canal and road cuts long before the period of the bulldozer. From the geological mapping of William Smith, a conventional, international color and notation scheme for rock types, based on their age and lithology, was subsequently developed. Geologists, including Smith, made important use of profiles or vertical cross sections for showing the relative altitude and the attitude of rocks (fig. 7.2), and thus the study of stratigraphy was greatly advanced. Profiles or cross sections of continents were employed by Baron (Friedrich Heinrich) Alexander von Humboldt in his monumental studies at approximately the same time.

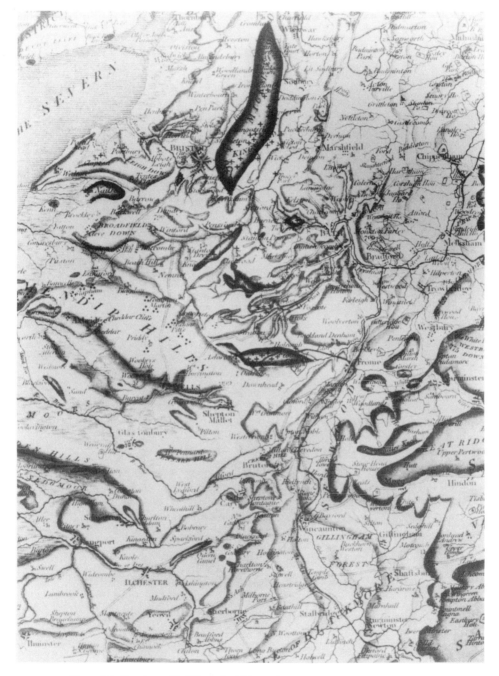

*Figure 7.1.* Portion of William Smith's geological map of England (1815).

Figure 7.2. Profile and cross section of the strata of England by William Smith (1815).

Two of the founders of modern geography, Humboldt (1769–1859) and Karl Ritter (1779–1859), were associated with the great cartographic publishing house of Justus Perthes in Gotha. In 1806 Ritter, who was especially interested in geographic education, published a generalized map of Europe in which he rendered particular altitude zones by means of bands of gray decreasing in intensity with elevation. Although not strictly a hypsometric tint map (apparently, as we have indicated, Dupain-Triel may be credited with this development), Ritter's map systematically employs the convention by which the higher the altitude, the lighter the tone used. The opposite method—that is, increasing intensity of tone with elevation—had been employed in a very generalized hypsometric map of the world by Johan August Zeune (1778–1853) published in 1804. Later the Austrians Franz von Hauslaub (1798–1883) and Karl Peuker (1859–1940) developed the conventional layer (color) tint system now most commonly employed.[4] This method, in which green is used for the lowest elevations followed by yellows at intermediate altitudes and brown at the highest peaks, was suggested by the humid European landscape and is not so suitable in all situations, especially mid- and low-latitude deserts.

Humboldt occupied a particularly influential position in the world of science in the first half of the nineteenth century. He was personally acquainted with many of the greatest thinkers of his time and became a statesman of science. Humboldt's early work was in mineralogy, chemistry, and botany, but later he embraced a wide range of knowledge, including physics, oceanography, and climatology, in his research. It was in the last of these fields of study that he made his most original contribution to cartography. But before devising his landmark map of isotherms (lines of equal average temperature), Humboldt journeyed extensively in the Americas with the French medical doctor and botanist Aimé Bonpland. They were given permission to travel and to make scientific observations in the Spanish empire and left La Coruña in 1799, when, ironically, Malaspina was languishing in jail in the same Spanish city. After his arrival in what is now Venezuela, Humboldt was able to confirm the existence of and to map the Casiquiare Channel, which connects Orinoco and Amazon drainage. Humboldt and Bonpland then went to Colombia and, by way of Magdalena River, on to Ecuador, where they climbed nearly to the summit of Chimborazo (6,267 meters), believed to be the highest ascent by humans up to that time. More mapping was accomplished before they left for Mexico (New Spain), where Humboldt was impressed by the quality of training received by topographical engineers in the Viceroyalty, half of which was soon to be taken over by the United States. It was in Mexico City, with access to great archival resources, that Humboldt compiled his "Map of New Spain," the best delineation of Central and North America (from 15 to 40 degrees north latitude, and from 90 to 115 degrees west longitude) made up to that time, based on astronomical observations.[5] On their way home to France in 1804 Humboldt and Bonpland visited President Jefferson in Washington, where, unwittingly, they provided cartographic intelligence on Mexico shortly to be of great value to the United States. Jefferson, following the Louisiana Purchase of 1803, had just dispatched Meriwether Lewis (1774–1809) and William Clark (1770–1838) overland to the Pacific Northwest in an exploration lasting from 1804 to 1806. Information from the sketch maps of these explorers was later incorporated into more general works such as Samuel Louis's "A Map of Lewis and Clark's Track, across the Western Portion of North America from the Mississippi to the Pacific Ocean," published in Philadelphia (1814).

Humboldt's map of New Spain was first published in France in 1811, but it is his isothermal map (fig. 7.3), resulting from his travels and observations on both sides of the Atlantic and published in 1817, that is his most enduring cartographic legacy.[6] He had noted that the average temperatures on the west sides of continents in the midlatitudes are milder, by and large, than those on the east coasts in the same latitudes, an idea that overthrew the classical idea of a strict zonality of climate according to latitude. To demonstrate this important concept, he drew a plane chart from the equator to 85 degrees north latitude and from 94 degrees west longitude eastward to 85 degrees east longitude, based on the prime meridian of Paris. Within this framework, every tenth parallel from 0 to 70 degrees north is drawn, but only three meridians. The average summer and winter temperatures of thirteen places are plotted in their geographical locations, but there are no coastlines or other geographical data. To this base Humboldt added isotherms, which reach their highest latitude at 8 degrees east longitude while the lows are at 80 degrees west and 116 degrees east longitude (the three drawn meridians). The curving isothermal bands contrast with the straight geographical parallels of the plane chart. Below the map Humboldt added a diagram to show the effect of altitude on the isotherms. Humboldt acknowledged his debt to Halley for development of the isoline concept, which was soon applied to other phenomena.

Among his other contributions to geography, which Humboldt expressed in beautiful maps and diagrams, was his theory of the vertical zonation of climate and vegetation. In other words, if a mountain—even one at the equator, such as Chimborazo—is high enough, all plant associations from the tropics to near the polar regions could be represented among the mountain's vegetation. He also greatly improved on Kircher's

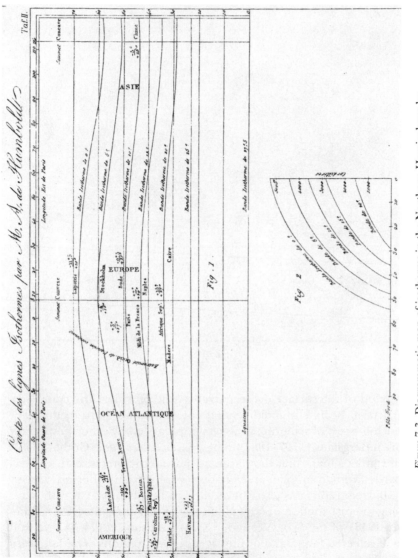

*Figure 7.3.* Diagrammatic map of isotherms in the Northern Hemisphere by Alexander von Humboldt (1817), with vertical zonation diagram below.

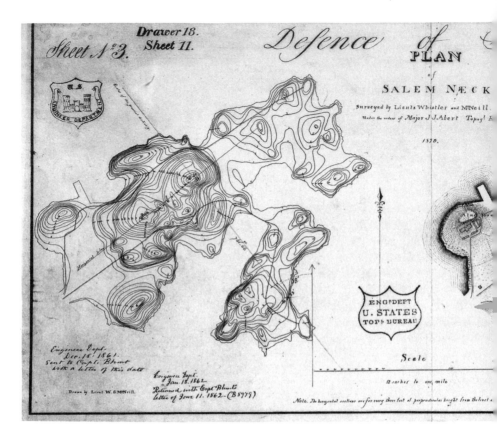

conception of the surface ocean currents, one of which, the cool Peru-
vian current, bears Humboldt's name on some atlas maps to this day.[7]
This rich harvest of scientific ideas was popularized by others, especially
Heinrich Berghaus (1797–1884) in his *Physikalischer Atlas* (Gotha, 1845),
which contains maps that show average barometric pressure at sea level
(isobars), average annual precipitation (isohyets), isotherms, and so
forth. Biogeography was also represented in Berghaus's atlas with maps
showing the distribution of crop plants, while in *The Physical Atlas* (Edinburgh, 1848) of the Scot Alexander Keith Johnston (1804–71) we find
maps of selected fauna and of phytogeographic regions. The latter are
derived from the research of the Dane Joakim Fredrick Schouw (1789–
1852), but all of these workers looked to Humboldt for inspiration.

In the previous chapter, we traced the development of terrain representation in Europe up to the systematization of the hachure by Leh-

*Figure 7.4.* Comparative maps of Salem, Massachusetts, illustrating the contour (*left*) and hachure (*right*) methods of terrain representation by George Whistler and William McNeill (1822).

mann and the use of contours for landforms by Dupain-Triel. The application of these methods to official American cartography is well illustrated by a map (fig. 7.4) produced by two early graduates of the United States Military Academy, Lieutenants George W. Whistler and William G. McNeill.[8] Whistler was the father of James Abbott McNeill Whistler (1834–1903), the painter, whose own training as a draftsman and map engraver in the Coast and Geodetic Survey was of great value in his career as an etcher. Figure 7.4 illustrates the comparative merits of the contour and the hachure methods. As indicated before, but most clearly shown here, the hachures are short strokes drawn downslope—the thicker the line, the steeper the slope. Absolute elevation can only be inferred from the expressive and graphic hachures. By contrast, the contours make it possible not only to read elevation but also to measure slopes within the limit of the contour interval (vertical distance between successive contour lines). However, the contour method is difficult for some map readers to understand because it is a more abstract form of symbolization. The battle for adoption of the contour in topographic mapping (a more quantitative method of terrain rendering, as opposed to the largely qualitative hachure) was not won as early as 1822, the date of Whistler and McNeill's plan of Salem. Actually, the contour method was not officially approved for the British Ordnance Survey maps until the mid-nineteenth century, and it was decades after this before most topographic sheets were contoured.[9]

The United States Military Academy at West Point, formally founded in 1802, was the only place in the country where Whistler and McNeill could have obtained the technical training requisite for their work as surveyors in the first decades of the nineteenth century. At this time the young republic still depended on Europe for scientific expertise. Thus the Swiss Ferdinand Rudolph Hassler (1770–1843) became professor of mathematics at West Point and was appointed superintendent of the U.S. Coast and Geodetic Survey when it was founded in 1807. Hassler and Joseph N. Nicollet (1786–1843), who was from Savoy but trained at the Paris Observatory, did much to influence mapping in the immediate post-Revolutionary War period. Nicollet undertook a large riparian survey which resulted in his "Map of the Hydrographic Basin of the Upper Mississippi River, from Astronomical and Barometrical Observations, Surveys, and Information, 1843."[10] Equally important was Nicollet's influence on two of the great cartographers of the American West, John Charles Frémont (1813–90) and William Hemsley Emory (1811–87).

Frémont had been assistant to Nicollet in the Upper Mississippi Survey, and with the help of the German cartographer Charles Preuss he later explored and produced reconnaissance maps of a large area of the United States, from the Rockies to California and Oregon, between 1842 and 1854. Like Frémont, Emory was an officer in the newly founded (1838) Corps of Topographical Engineers, U.S. Army.[11] In this capacity, after having brought Nicollet's map to publication form, Emory embarked on a great survey of the arid Southwest of the United States. Using methods employed earlier by Nicollet, Emory made a map, "Military Reconnaissance of the Arkansas, Rio del Norte [Rio Grande] and Rio Gila" (1847). Following the 1848 Treaty of Guadalupe Hidalgo, Emory went on to demarcate and map the three thousand kilometer U.S.-Mexico boundary, which occupied him until 1857. The highest peak in the Big Bend Country of Texas is named Mount Emory (2,370 meters). Both Emory and Frémont were generals in the American Civil War (1861–65) and were among those who espoused the extension of the

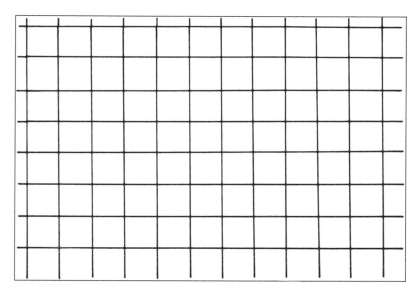

*Figure 7.5.* Map comparing basic cadastral survey units:
sections of the U.S. Public Land Survey above,
and metes and bounds divisions below.
Both embrace areas of one hundred square miles
in Ohio (ca. 1820).

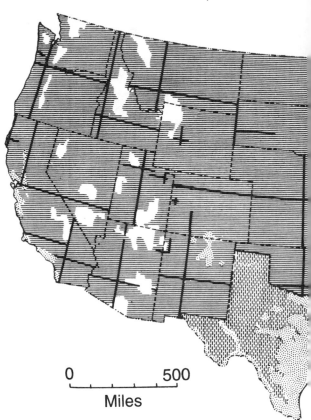

0 _____ 500
Miles

railroads from the East to the Pacific coast.[12] A map that more particularly addresses this problem is that of Lieutenant Gouverneur K. Warren, which accompanies the *Pacific Railroad Reports* of 1857. In compiling this map of the trans-Mississippi West, Warren could utilize source maps of the area made over the preceding half-century and earlier. We will treat transportation mapping in the nineteenth century subsequently, but we will now deal with cadastral mapping, especially in the United States.

Following the Revolutionary War and the ratification of the Articles of Confederation, the new republic had fallen heir to a large public domain, which, as we have seen, was in time to be expanded to include a vast tract of land from the Atlantic to the Pacific—the conterminous states of the United States. The subdivision and settlement of this area was a major problem. For these purposes, a committee was set up under the chairmanship of Thomas Jefferson, and, as a result of its report, Congress enacted the famous Land Ordinance of 1785. Elements of this legislation that are of particular concern here are (1) survey prior to settle-

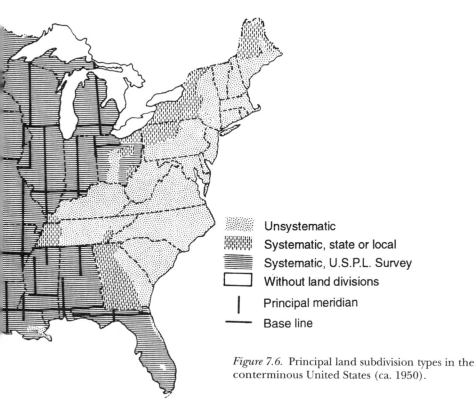

Unsystematic
Systematic, state or local
Systematic, U.S.P.L. Survey
Without land divisions
Principal meridian
Base line

*Figure 7.6.* Principal land subdivision types in the conterminous United States (ca. 1950).

ment; (2) orientation of survey lines; (3) the township unit; and (4) the section. To implement this systematic cadastral survey, Thomas Hutchins, who had served as a surveyor under George Washington, was appointed geographer of the United States, or first officer in charge of government surveys. With the magnetic compass and chain as the principal surveying instruments, the United States Public Land Survey (USPLS) was begun in eastern Ohio.[13] As surveys progressed westward, some modifications were made in the system, but in general unsettled land was subdivided into square-mile sections. These were organized into townships of thirty-six sections each (six by six sections), with blocks of townships controlled by principal meridians (N-S) and principal bases (E-W). Sections were later subdivided into properties, resulting in a network of fundamental survey lines oriented predominantly in cardinal directions. This system stands in sharp contrast to the metes and bounds survey of the eastern seaboard of the United States (fig. 7.5) and of most of the older settled parts of the world (fig. 7.6). There are some irregularities in the USPLS, owing to correction lines, because a square grid is fitted to converging meridians and because compass bearings are not always accurate. But, by and large, the survey is remarkably rectilinear.

At first the pace of cadastral surveying in the public domain was slow, but later, as settlers clamored for land, it increased in speed. By the time of the American Civil War, a large part of the humid lands of the eastern half of the United States had been subdivided by surveyors employed by the General Land Office. During the war many of these surveyors enlisted and were employed in a variety of military mapping activities. On the restoration of peace, cadastral surveying again occupied the attention of many workers. In addition to the original survey plots, compiled maps of townships and counties were required for a variety of legal and administrative purposes. The result was a remarkable number of published and unpublished, official and unofficial, cadastral maps of the agricultural areas of the United States.[14] Production of these maps proved to be a lucrative proposition for publishers, who in the 1860s began to develop county atlases.[15] A phenomenal number of these relatively expensive, hard-covered volumes were printed and sold in the following decades. Before the end of the century, some counties in the Corn Belt could boast of ten editions of an atlas, while half of the counties of the United States were not even covered by such atlases. Figure 7.7 is a typical page from a county atlas of the 1870s in the area of systematic surveys. It shows the boundaries of all the properties in a surveyed township, which in this case, as frequently, was also an administrative (political) township. The property lines and roads commonly extend north-south and east-west without reference to rivers or, for that matter, to the form of the land. (The USPLS has been characterized as "the triumph of geometry over geography.") As is typical of older cadastral maps, no relief is shown, but the size and ownership of properties and the location of dwellings are emphasized. Engraving, and later the cheaper process of wax engraving, was used to reproduce the maps. County atlases were usually sold on a subscription basis, and they contain pictures of patrons, their families, and the establishments of those willing to pay for this privilege. Typically, the pictures are lithographs.

In his great American novel, *Raintree County*, in which the county atlas is a central motif, the author Ross Lockridge Jr. wrote:

> The *Atlas* was remarkable for its illustrations, fullpage lithographs of the New Court House, Freehaven's leading hotel, the south side of the Square; and half a hundred pictures—some fullpage and some two and four to a page—of Raintree County homes, mostly farms.

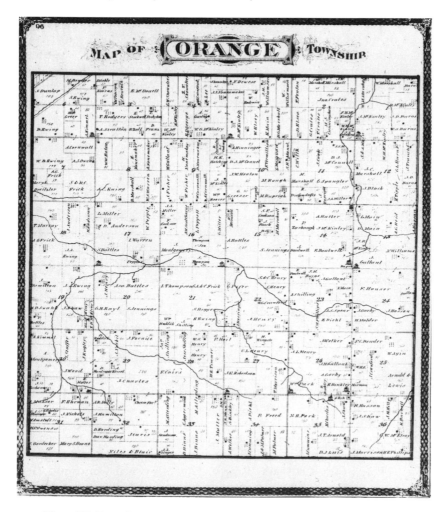

*Figure 7.7.* Page from a county atlas of Ohio showing property boundaries in the thirty-six sections of a surveyed township, which, in this case, was also a civil township (1877).

Lockridge continues in this vein for several paragraphs, which suggests that the county atlas of the United States well expresses the spirit of the free-enterprise, agricultural society that gave it birth.[16] It is a valuable, though neglected, source of geographical information of the period.[17]

A cartographic form as important for nineteenth-century urban places as the cadastral map was in rural areas was the fire insurance and underwriter's map. Although developed in England in the last years of the eighteenth century, such maps reached their zenith in the United States in the second half of the nineteenth century. These remarkably detailed maps show the fire-resistive character of buildings and other types of information. The need for constant revision led to the development of other means for recording these data in the twentieth century, so that they have largely been superseded and are now of historical interest only.[18] However, more traditional urban mapping, with its origins in the beginnings of cartography, flourished in the nineteenth century, as it does today. Two principal forms of urban maps can be recognized: the plan view, represented earlier by the multisheet map of London (1742) by John Rocque (fl. 1732–64), among many other examples; and the bird's-eye or oblique view, which has its precedents in the masterly six-sheet, woodcut view of Venice (1500) by Jacopo de Barbari (ca. 1440–1515), engravings in the *Civitates orbis terrarum* of Braun and Hogenberg (see fig. 5.12 in chap. 5), and the map of Amsterdam (1544) by Cornelis Anthonisz (1499–1557), to name only a few. Two types of the latter genre can be recognized: the isometric (see fig. 3.5 in chap. 3) and the perspective view.[19] Engraved and lithographic representations of literally thousands of settlements in plan view and in both types of oblique view were made in the nineteenth century. Particularly interesting in this field, because of the combination of plans with vignettes of city views, are maps published by the (British) Society for the Diffusion of Useful Knowledge between 1830 and 1843. Maps of the principal cities of the world, including such places as Calcutta, were the subjects of the steel engravings of the SDUK.[20] Steel engraving is a particularly difficult medium for cartography, but a single plate will yield many more copies than softer metal such as copper. But the great step forward during this period in printing was lithography, developed by Aloys C. Senefelder (1771–1834) in the last years of the eighteenth century and first used for maps in the early nineteenth century. This process allows the production of continuous tonal variation or shading, which is of great utility in cartography. It was used for some multiple-color and shaded-relief maps in the nineteenth century but came to its full realization through its marriage with photography, mostly a twentieth-century development.[21]

We have encountered route maps from antiquity, especially Roman maps, and it has been said that there was no improvement in the speed of land travel in Europe from Caesar to Napoleon. All of this changed with the advent of the railroad in the early years of the nineteenth century. At first locomotives were confined to mines, but in 1825 a railroad was laid out between two discrete places, Stockton and Darlington in northern England, a distance of about fifteen miles. With this, passenger and freight lines began, and the railroad age was initiated. Soon railroads extended between metropolitan centers of the industrialized world, even internationally. This gave rise to a new field of cartography when railroad maps became an important adjunct to other forms of promotion, as well as being informative to those who would use the new mode of transportation.[22] In some of the larger countries of the world, railroads—and, by extension, their representation on maps—became of surpassing importance. Thus India was bound by bands of steel, Canada became a nation, and the United States reached its "manifest destiny" with the help of the railroads. Railroad maps, like political maps, are not usually considered the most visually exciting form of cartography, but to those who need to consult them, they provide data of infinite value. Like many other map forms, they were often regarded as ephemera, and thus all but a few copies have perished. Fortunately there are some collectors of such materials, so we have a record of the significance of railroad maps in the nineteenth century and beyond. Lithography was frequently used in their reproduction, often on paper of poor quality (another factor leading to their destruction), and later wax engraving was used.

One unanticipated result of the extension of railroads in great east-west directions, as in the United States and Canada (or west-east, as in the case of Russia), was the urgent need for the establishment of a single prime meridian and of time zones; the desirability of the former had been long understood by sailors. After a series of preliminary meetings, these problems were resolved at the International Meridian Conference held in 1884 in Washington, D.C., when Greenwich, England, was approved (though with much dissension) as the global prime meridian and the center of the first of twenty-four world time zones.[23] On maps of these zones, a clock face with an appropriate time would sometimes appear in the middle of a particular zone, as on the map "The Mexican Central Railway, 1890," which has four clocks for the United States and one for Mexico (fig. 7.8). More often the time zones on regional and world maps are differentiated by bands of shading or color. Such maps, on which time-zone boundaries jog around political units and on which the international date line (180 degrees longitude) is featured, became and remain popular. The success of the railroads was so great that some toll and wagon roads that had previously been quite serviceable fell into disrepair, and on many general atlas maps of the nineteenth century, railroads (and canals), but not highways, were often shown. It was only in the twentieth century, after automobiles became a favorite mode of land transportation, that roads were greatly improved. With this came the modern road map, which will be discussed in the next chapter.

A development somewhat parallel to the railway was the application of steam power to ships. However, it took some time before most ocean-

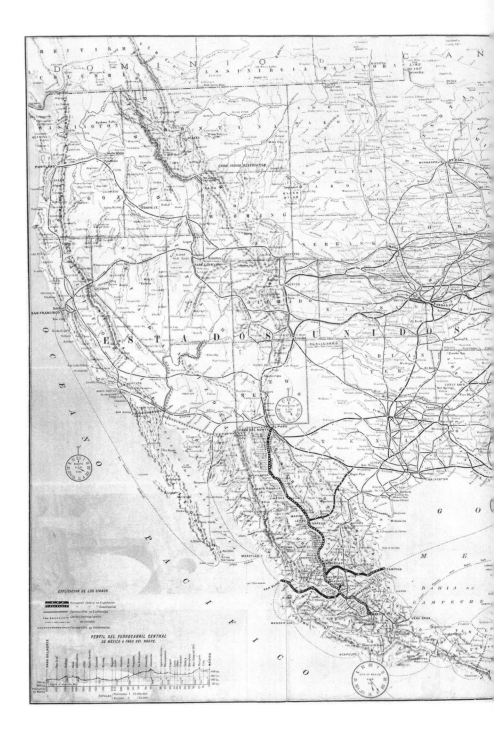

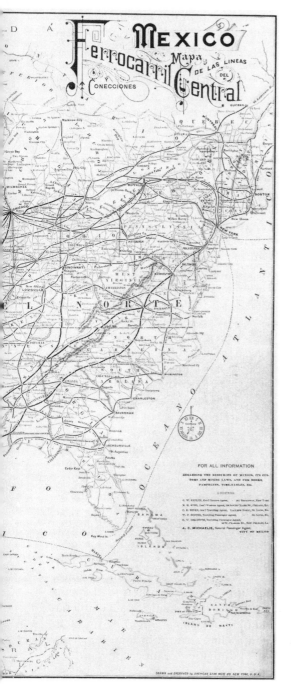

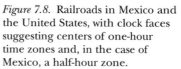

*Figure 7.8.* Railroads in Mexico and the United States, with clock faces suggesting centers of one-hour time zones and, in the case of Mexico, a half-hour zone.

going vessels were equipped with engines, and in this development the nineteenth century was a time of transition. Britain dominated the seas during this period, and charts from the Hydrographic Office of the Royal Navy were to be found aboard naval and merchant ships of many nations—one reason, but not the only one, for the establishment of Greenwich as the international prime meridian. But in terms of cartography, other nations were pioneering new methods. In particular, marine charts of greater sophistication and usefulness to sailors than any produced up to this date were prepared under the direction of Matthew Fontaine Maury (1806–73) in the United States.[24] Maury, who entered the United States Navy in 1825, became interested in a variety of navigational problems. His writings on scientific subjects led to his appointment in 1842 as superintendent of the Depot of Charts and Instruments, whose functions were later divided between the United States National Observatory and the Hydrographic Office. It was during this period as superintendent that Maury produced his celebrated wind and current charts. At first he used old log books as a source of data, but later he requested masters of naval and commercial vessels to send in reports of their voyages on forms prepared specially for this purpose. Generalizing from the great mass of information he received, Maury made recommendations for the quickest passages between various ports according to winds, currents, and other factors. These tracks, which in some instances deviated considerably from great-circle routes (those of the shortest distance between two points on the globe), led to savings of days and even weeks on long voyages. A portion of an engraved chart by Maury of the Pacific Ocean off California is featured in figure 7.9; the accompanying legend, or key, indicates the symbols used on such charts.

Maury prepared maps showing various facets of the physical characteristics of the world's oceans, but his wind and current charts, as illustrated, are best known. He received international acclaim for his endeavors, becoming the American most decorated by foreign governments up to his time. As suggested, Maury's recommendations came at a time when sail was soon be to superseded by steam, which liberated shipping, to a large extent, from dependence on winds and currents, but his ideas were original and he served as a "one-person computer" before the computer age. As others had done earlier, Maury made deep-sea soundings, but not enough data could be collected by methods then available to permit a satisfactory map of the ocean floor to be made. Maury summarized his ideas in a book, *The Physical Geography of the Seas* (1855), which, along with his other writings and charts, has assured him an imperishable place as one of the founders of systematic oceanography and the

"Pathfinder of the Seas." In the Civil War he served his native South as a naval officer but, unlike General Robert E. Lee, suffered as a result of this during the Reconstruction. Prior to the war, Charles Wilkes (1798–1877) also commanded extensive marine explorations on behalf of the United States Navy, from which his world chart of 1842 resulted.

The weather map, on which a variety of meteorological phenomena are plotted together, had its origins in the nineteenth century. Maps made in the 1840s by the American Elias Loomis (1810–99) are among the earliest weather maps, but ideas for these came from several sources, including the wind velocity scale devised by Sir Francis Beaufort (1774–1857), hydrographer of the Royal Navy. In times of peace this service also engaged in sponsoring research, as when Charles Darwin (1809–92) was attached as a naturalist to HMS *Beagle* on a coastal surveying assignment. It was on this voyage that Darwin made his momentous discoveries leading to the theory of evolution. But charts made by official charting and mapping organizations should not always be relied on, as indicated by the persistent inclusion of some mythical islands that had a long life before they were finally removed from the cartographic record.[25]

We have discussed the origins of thematic mapping and noted the contributions to this field by such well-known scientists as Halley and Humboldt. Like these men, Henry Drury Harness (1804–83) was a thematic cartographer who, at least during his lifetime, was better known in other connections.[26] Harness had a distinguished career as an administrator, for which he eventually received a knighthood. He graduated from the Royal Military Academy at Woolwich and while still a lieutenant was employed by the Irish Railway Commission. It was in this service that he supervised the construction of three maps that were published by the commission in 1837 and that have now established his reputation as an important cartographic innovator. Harness could look to the example of William Playfair (1759–1823), who had devised graphs of economic production and diagrams of relative areas and population by superimposed squares in his *Commercial and Political Atlas* (1786). However, Harness went beyond Playfair in actually mapping statistical data. In his population map of Ireland (fig. 7.10), Harness used the dasymetric method of showing classes of demographic density in rural areas.[27] In this technique, the areal symbols, which in Harness's maps are reproduced by the aquatint process, are not confined by administrative boundaries but cover areas of homogeneity within specified limits. They differ from areas defined by isopleths (such as Humboldt's isotherms) in that higher and lower values can be adjacent to one another without the necessity of intermediate values in between. On the same map, urban centers are

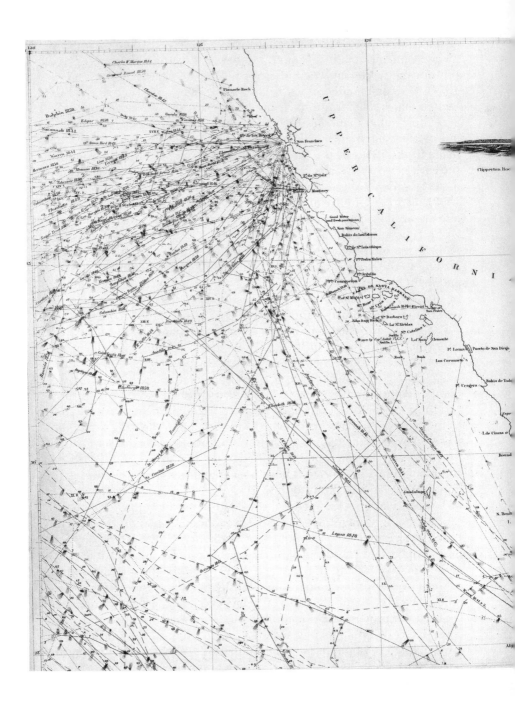

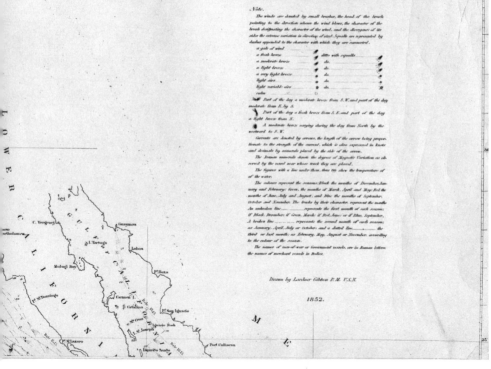

*Figure 7.9.* Chart showing winds and currents off the California coast by Matthew F. Maury (1852).

*Figure 7.10.* Dasymetric map of the population of Ireland by Henry D. Harness (1837).

shown by filled circles proportional in size to the population of the place. This same device is used for urban places in Harness's traffic-flow map (fig. 7.11), which also shows flow lines proportional to the relative quantities of traffic in different directions. A third map (not illustrated)

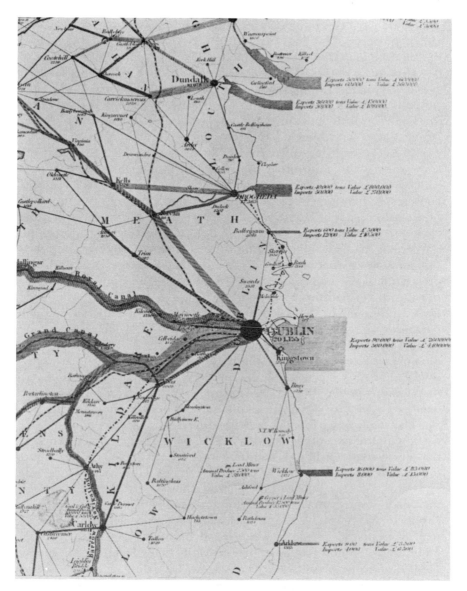

*Figure 7.11.* Portion of traffic-flow map of Ireland utilizing quantitative flow lines and graduated circles by Henry D. Harness (1837).

employs similar symbols to show the number of passengers in different directions by regular public conveyances, and again the relative sizes of cities are indicated by proportional circles. In his maps for the Railway Commission, Harness seems to have originated all of these elements:

graduated circles (for cities) according to population, density symbols (for population), flow lines (for traffic and conveyances), and the dasymetric technique (for population in rural areas).

However, it remained for the Danish naval officer Nils Frederik Ravn (1826–1910) to use isolines for a social or cultural phenomenon (as opposed to a strictly physical one) in his population density map of Denmark (1857). Such lines are known generically as *isopleths*—that is, lines connecting points assumed to have equal value—and, in this case, specifically as *isodems* (see appendix B). Ravn's work is later than that of some others who were using different methods to map population, which was possible as soon as demographic data became available from censuses. At first such maps were spatially simple, such as the dot map of the *départements* of France made by Frère de Montizon in 1830, with each dot representing ten thousand people and the requisite number of dots evenly spaced within the administrative units. Apparently this is the first map on which uniform symbols were used to represent a number greater than one (uniform city symbols do not represent the same numbers). Another statistical device was used for a map of France compiled by A. D'Angeville in 1831, with the density of the population within *départements* represented by seven tonal shades, from white (highest) to darkest (lowest), the opposite of the scale normally used today. D'Angeville's map is an example of the simple choropleth technique of quantitative mapping. By this method, the quantity of the phenomenon being mapped is considered in relation to the size of the statistical areas (for example, counties), density is computed, and density categories are set up. The whole of a particular area is colored (or shaded) uniformly according to the category into which it falls. Obviously in its lack of internal differentiation within a statistical unit the choropleth differs from the dasymetric method used by Harness, but it is more "objective."

Soon a large number of population characteristics were mapped, including poverty, crime, sanitation, and disease. Understandably, medical doctors were involved in this last type of activity, which included a map of hernia in France by Joseph-François Malgaine (1840); one of cretinism in Canton Aargau, Switzerland, by Ernst Heinrich Michaelis (1843); and one of influenza in Glasgow, Scotland, by Robert Perry (1844). But the greatest number of maps of disease at this time were of Asiatic cholera, which appeared in Europe in the early 1830s. General maps of this disease were made by Dr. Robert Baker of Leeds (1832) and J. N. C. Rothenburg of Hamburg (1836). A map showing individual cases of cholera in Exeter was made by Dr. Thomas Shapter in 1849. This indicates deaths in the three years 1832, 1833, and 1834, each year's deaths differentiated by a uniform symbol (dot, cross, open circle) and

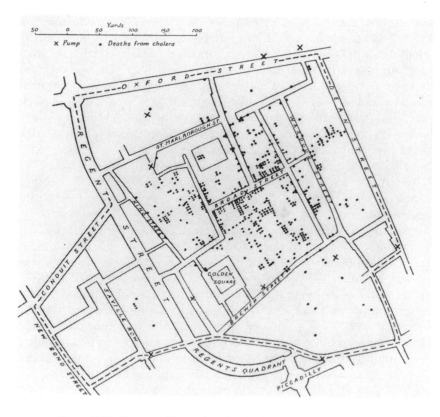

*Figure 7.12.* Dot map illustrating deaths from cholera in London by
Dr. John Snow (1855). Uniform dots are used in this reconstruction
rather than the small rectangles of the original.

numbers keyed to associated features (soup kitchens, drugstores, and
places where clothes were destroyed). However, the best-known map of
cholera was made by Dr. John Snow in 1855 of individual deaths from
the disease in a small district of London, each demise represented by a
uniform symbol—in this case, a small rectangle (fig. 7.12). By his map
Dr. Snow discovered that "the incidence of cholera was only among persons who drank from the Broad Street pump."[28] At Snow's request, the
handle of this pump was removed, and new cases ceased almost at once.
Further investigation showed that the well on Broad Street was near a
sewer.

Dr. Snow was largely responsible for demonstrating the water-borne
origin of cholera, and he published his findings in his book, *On the Mode
of Communicating Cholera* (London, 1858). He is honored with a public
house named for him near the site of the offending pump. Dr. Snow

also made a map showing districts served by two water companies on which it is revealed that, understandably, areas found to be receiving relatively unpolluted water had a dramatically lower death rate than those supplied with water in contact with raw sewage. Dr. Snow's maps illustrate the highest use of cartography: to find out by mapping that which cannot be discovered by other means or, at least, not with as much precision. Apparently the first map in a general atlas to show the distribution of human diseases was in Berghaus's landmark physical atlas, noted previously. In the United States, Richard Swainson Fisher, M.D., contributed descriptions "Geographical, Statistical, and Historical" to *Colton's Atlas of the World* by George W. Colton (New York, 1856), another popular compilation, but these descriptions do not have to do with health problems.

The maps of Drs. Shapter and Snow differ in a very important respect from most modern dot maps: the value of the uniform symbol is one. In the population map of de Montizon (1830) we have seen an example of such symbols representing more than one—obviously a higher order of generalization, but not in that case placed in "correct" geographical concentrations. Gradually the idea took hold of uniform symbols representing a number larger than one (ten thousand in the case of the map of de Montizon) combined with correct geographical distribution within the statistical unit, a task not accomplished without overcoming difficult cartographic and geographical data problems. Figure 7.13 illustrates the four principal methods used in thematic mapping today, all of which were devised in the nineteenth century. August Petermann (1822–78) was one of those who worked to extend and popularize statistical/geographical cartography through his publications. The mapping of the distribution of crime was undertaken by André Michel Guerry (1802–66) and Adrien Balbi in their *Statistique morale de l'Angleterre comperée avec de la France* (1864) using choropleths. The flow line, devised by Harness, was popularized by Charles Minard (1781–1870) in his *cartes figuratives*.[29] In his numerous diagrammatic cartographic presentations, Minard was willing to alter spatial relationships (anticipating the modern cartogram) to suit his purposes, and he dealt quantitatively with various facets of economic, social, and historical geography. Minard also used the graduated circle and sectored some of these to make so-called "pie graphs"; these have since become popular cartostatistical devices. All of the common techniques of thematic cartography in use today had been developed by 1865.

A great interest in statistics—and, by extension, statistical mapping—was evinced by geographical societies in the nineteenth century, as when the eminent statistician Sir Francis Galton (1822–1911) was employed by the Royal Geographical Society of London. This was one of

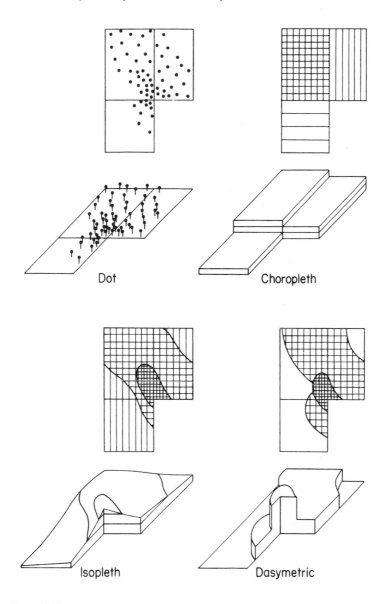

*Figure 7.13.* Principal methods of statistical mapping illustrated in two- and simulated three-dimensional form.

several such societies founded in the early nineteenth century particularly to solve geographical problems, especially those concerned with the interiors of the continents—Africa, Asia, North and South America, and Australia.[30] Many geographical discoveries concerning the height and alignment of mountain ranges, the relative aridity and extent of deserts, the sources and courses of rivers, and so forth were recorded on sketch or reconnaissance maps, soon to become part of the more general cartographic record as published maps. We have mentioned the primacy of Italy in the foundation of a geographical (cartographical) society through Coronelli, and others were founded, especially in Germany, in the eighteenth century. The three oldest European geographical societies north of the Alps that are still in operation—those of Paris, Berlin, and London—are all attributable, indirectly, to the voyages of James Cook. The first of these societies, that of Paris, founded in 1821, was inspired by the success of British naval officers such as Samuel Wallis (1728–95), the European discoverer of Tahiti, and, of course, Cook. The society prompted the French to expedite their own explorations and to take a greater interest in the Pacific Ocean. The second society—that of Berlin, founded in 1828—owes much to Humboldt, who had been influenced by Johann Georg Adam Forster (1754–94), who had accompanied his father, the naturalist Johann Reinhold Forster (1792–98), on Cook's second Pacific voyage. Like others at the time, Humboldt thought the Andes were the highest mountains on earth and was disappointed when measurements by the Survey of India, in which Indians took an active part, revealed that Chomolungma (later Mount Everest) was higher. The Royal Geographical Society, founded in 1831, grew out of the African Association, founded in 1788, in which Sir Joseph Banks (1734–1820) was a leading spirit. Banks was a botanist on Cook's first Pacific voyage, and although not usually thought of as a cartographer, he did engage in the mapping of the Society Islands and became a patron of cartographers as president of the Royal Society from 1778 to 1820.

During the nineteenth century the cartographic output of geographical societies including maps published in their journals and special projects, is impressive. The names of those who provided raw material for this work are numerous and well known, and only a few will be given by way of illustration: Mungo Park (1771–1806), David Livingstone (1813–73), Richard F. Burton (1821–90), John Hanning Speke (1827–64), and Henry Morton Stanley (1841–1904) in Africa; Nikolay M. Przhevalsky (1839–88) and Auguste Pavie (1847–1925) in Asia; (Friedrich Wilhelm) Ludwig Leichhardt (1813–ca. 1848) and John M. Stuart (1815–66) in Australia; and Samuel Hearne (1745–92), Alexander Mackenzie (1764–1820)—who crossed the North American continent a decade before Lewis and Clark—and Simon Fraser (1776–1862) in Canada. Grants were given and prizes awarded for geographical "firsts" by the societies, which maintained cartographic staffs to transform explorers' rough sketch maps into finished renderings for publication.

The quest for geographical priority became a nationalistic contest in the nineteenth century after the Napoleonic Wars and resulted in much reconnaissance charting and mapping. Thus the discovery in 1840 of the previously postulated South Magnetic Pole, claimed by both the Frenchman Jules-Sébastien-César Dumont d'Urville (1790–1842) and the American Charles Wilkes, led to an improved cartographic delineation of part of the coast of Antarctica. The modern search for the source of the (White) Nile (the longest river in the world), which had baffled humans since antiquity, was initiated by two Englishmen on leave from the Indian Army, Burton and Speke. Its discovery by Speke and the Scot James Augustus Grant (1827–92) had a profound effect on the mapping of interior Africa in the 1850s and 1860s and later, under other explorers. Baron Nils Adolf Erik Nordenskiöld (1832–1901), a distinguished scientist/cartographer, navigated the Northeast Passage (north of Eurasia) from 1878–1879 in his ship *Vega*. This Swedish Finn spent his later years studying the history of cartography and assembled a great collection of maps and associated geographical materials, now in Helsinki. We have already discussed the epic work of Humboldt in Latin America and the mapping endeavors of Frémont, Emory, and Warren in the trans-Mississippi West of the United States. The reconnaissance maps of the latter area had been anticipated by the route maps of the pathfinders and mountain men, who in turn were indebted to Indians (Amerindians) for data. The mountain man par excellence was Jedediah Strong Smith (1798–1831), who traversed and mapped a route from the Rockies via Great Salt Lake across the "American Desert" to Mexico's Alta California and back (1826–30).[31]

It was in California that the postreconnaissance phase of the mapping of the American West began with the appointment of Josiah Dwight Whitney (1819–96) as head of the Geographical Survey of the new state in 1860 (fig. 7.14).[32] During the next fourteen years, Whitney and his assistants conducted a rigorous scientific survey that included measuring and mapping the highest (Mount Whitney, 4,418 meters) and lowest (Death Valley, −86 meters) points on the land area of the contiguous United States and exploring "the incomparable valley" of Yosemite. Also important, this survey proved to be a training ground for some of the scientists who were soon to be engaged in federal surveys, such as Clarence King (1842–1901).

Following the Civil War, four important surveys were undertaken in the area between the Great Plains/Rocky Mountains and the Pacific Ocean, each of which takes its name from its leader: King, Wheeler, Hayden, and Powell. In these surveys, a wide range of scientific problems was investigated, and mapping formed a very important component of

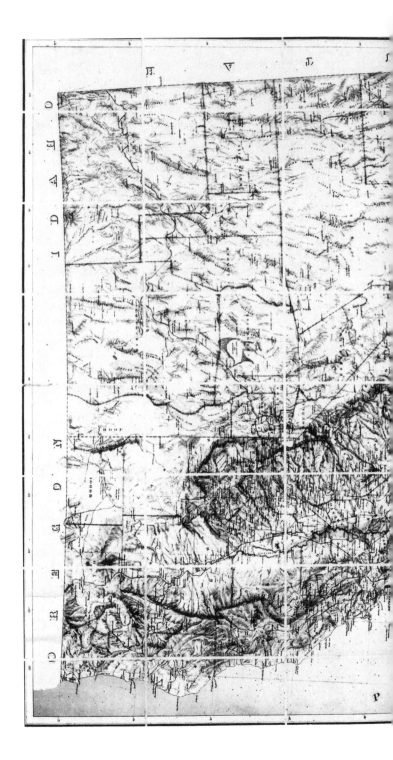

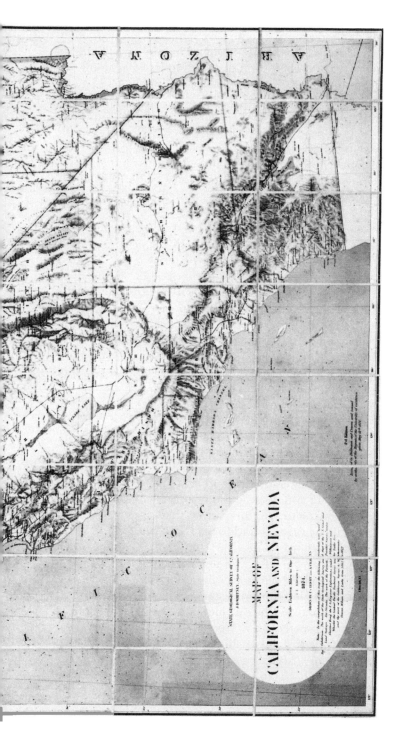

*Figure 7.14.* Shaded-relief lithographic map of California and Nevada made under the direction of J. D. Whitney, State Geological Survey of California (1874).

all the expeditions. Without going into any great detail, we can say that King's party, which included the photographer Timothy H. O'Sullivan (1840–82), examined a swath about one hundred miles wide along the fortieth parallel from the Great Plains to California between 1867 and 1877. George Montagne Wheeler (1842–1905) and his party, which included the geomorphologist Grove Karl Gilbert (1843–1918), surveyed the area west of the one hundredth meridian between 1872 and 1879. At roughly the same time, Ferdinand Vandeveer Hayden (1829–87) led a party that focused on Yellowstone and included the artist and topographer William H. Holmes (1846–1933) and the photographer William H. Jackson (1843–1942). Meanwhile, John Wesley Powell (1834–1902) and his party explored widely and traversed the Grand Canyon by boat in 1869. Among the achievements of these surveys were topographic mapping, including, in some areas, contouring over a larger region than previously undertaken in the United States; a pictorial record created through both drawing and photography; and ethnographic studies.[33] As a result of these activities, the United States Geological Survey was founded in 1879 with King as director. He resigned in 1881 and was succeeded by Powell, who is now regarded as the real founder of the USGS, which became the official, primary mapping agency of the United States. A further result of these western surveys was the creation of National Parks such as Yellowstone, which is about as large as the state of Rhode Island. The delineation of these protected areas, among the chief glories of the United States and Canada, resulted in maps for scientific, promotional, recreational, and later ecological purposes.

As in the preceding centuries, a number of nineteenth-century mathematicians advanced the study of map projections.[34] Of the many developments in this field during this period, only those in use today will be emphasized; salient characteristics and principal uses of selected projections are contained in appendix A. Some entirely new projections were invented in the nineteenth century, while new cases were developed of existing ones and new uses found for others. Thus in 1803 Christian Gottlieb Theophil Reichard (1758–1831) of Weimar employed the gnomonic projection, invented in antiquity and then used only for astronomical purposes, for terrestrial maps. On the gnomonic projection, a straight line is a great circle, the shortest distance between two points on the globe. This attribute makes the projection valuable for plotting routes that may be approximated on a Mercator chart with a series of straight line segments (rhumb lines), which can then be navigated by compass. The two projections, Mercator and gnomonic, are used in concert for navigation, although electronic/satellite positioning in superseding this method in the twentieth century. An equal-area, oval world projection was launched in 1805 by Karl Brandan Mollweide (1774–1825) in which all parallels are straight lines and meridians are semiellipses (see fig. 9.11c in chap. 9). This arrangement of the earth grid is usually known today as the homolographic projection. Another equal-area projection invented at about the same time as the elliptical Mollweide is the conic projection of Heinrich Christian Albers (1773–1833). The Albers projection, as well as being equivalent (equal area), has correct scale along its two standard parallels (see fig. 9.10e in chap. 9).

We have mentioned how the Swiss Hassler came to the United States and became superintendent of the Coast and Geodetic Survey. Hassler invented the polyconic projection (1820), in which a series of nonconcentric, standard parallels (rather than one or two) is used to reduce distortion. This projection was used for topographic sheets and atlas maps and later, with an important modification, for the International Map of the World. Arguably the most famous mathematician of the nineteenth century was Carl Friedrich Gauss (1777–1855). Among his many accomplishments was the ellipsoidal transverse Mercator projection, in which, as its name suggests, Gauss projected an ellipsoid on a sphere to produce a conformal projection (unlike Lambert, who had earlier used the sphere itself for such a transformation). The transverse property, as in the Lambert, Gauss, and other projections, is accomplished by rotating the grid through 90 degrees so that the poles, rather than the equator, are central in the projection. (The term *oblique* is used for a shift of less than 90 degrees from the normal position, as used on astronomical charts from the early eighteenth century). To Gauss we owe the term *conformality*, indicating correct shape around a point.

In the second half of the nineteenth century the study of map projections was continued by mathematicians and gifted amateurs, one of the latter being the Scottish-born clergyman James Gall (1808–95). He was especially interested in cylindrical projections, of which three are attributable to him as outlined in a 1885 publication: the Gall orthographic (equal-area); the Gall isographic (equirectangular); and the Gall stereographic (with the point of perspective on the equator opposite a given meridian). Each of these projections had standard parallels at 45 degrees north and 45 degrees south latitude. In order to reduce high-latitude exaggerations, as in Mercator's projection, Gall conceived the idea to have the cylinder intersect the sphere as shown on figure 9.10d in chapter 9. The Gall cylindrical equal-area projection has surfaced recently as the "Peters" projection, named by Arno Peters (b. 1916), who claims to have invented it in 1967! Among the other professional scientists who worked on projections in this period was the Victorian Astronomer Royal, Sir George Biddell Airy (1801–92). Airy produced a minimum-error azimuthal projection for one hemisphere that has no positive qualities (conformality, equivalence, and so on) but is a good compromise, and it became the bellwether of a large number of such arrangements of the earth grid, which continue to be proposed to the present day. Many other projections, some more decorative than useful, were produced in the nineteenth century, including the star projection by Hermann Berghaus (1828–90) in 1879 used later (1911) as the insignia of the Association of American Geographers. A method of assessing the distortion on map projections by means of an "indicatrix" was devised by Nicolas Auguste Tissot (1824–95), who published it in 1881.

Another development that found increasing use at this time was the *block diagram*. A block diagram is an oblique or three-quarter (isometric, or one- or two-point perspective) view of a slab of the earth's crust.[35] Surface features are usually portrayed on the upper portion of the block, while on the sides the underlying rock structures are shown. The block diagram thus unites the horizontal dimension of geography with the vertical element of geology, a device of particular value in geomorphology (the systematic study of landforms). Geological profiles were prepared by William Smith and by Humboldt, as indicated earlier, and Sir Charles Lyell (1797–1875) used these devices in rather naturalistic settings, with a suggestion of the associated surface features. Block diagrams were made by mining engineers to show mines, quarries, and caves, especially in Europe. In the hands of American geomorphologists Grove K. Gilbert (1843–1918) and especially William Morris Davis (1850–1934), the block diagram with its conventional rock symbols on the sides of the slab was fully developed (fig. 7.15). If well drawn, the block diagram can be a geometrically consistent view of the earth. Davis, whose ideas on earth science have now been largely superseded, nevertheless had a fine understanding of graphical means of expression and used series of block diagrams to illustrate temporal changes in the landscape, foreshadowing the truly animated diagram and map of later times.

Photographic processes suitable for map reproduction were developed in the second half of the nineteenth century, but photography was also to provide an improved data source—aerial photography (using the balloon as a platform in the nineteenth century and, after 1910, the airplane)—which has recently revolutionized mapping to as great an extent as did printing in the fifteenth century. These potentials were not realized until the twentieth century. The engraved topographic map based on instrumental field surveys continued to hold its own in the nineteenth century, so aerial photogrammetric surveys will be considered in our discussion of modern cartography.

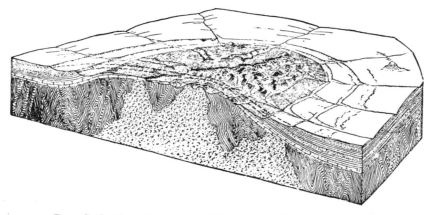

*Figure 7.15.* Block diagram by William Morris Davis illustrating the relationship between surface features and subsurface structures of the Black Hills, South Dakota (ca. 1898).

We shall also have occasion to refer in the next chapter to national atlases, the earliest of which has been revised and is available today: that of Finland, first published in 1899. This was sponsored by the Finnish Geographical Society, one of a number of such organizations founded in various countries in the nineteenth century that did much to promote cartography as well as geography.[36] Actually the *Statistical Atlas of the United States* compiled by Francis A. Walker predates the *Atlas of Finland* by some fifteen years. Walker's volume employs interesting symbolization and color to show a variety of physical and cultural distributions in the United States based on the 1870 census. It was compiled and published under the authority of the U.S. Congress in 1874 but, unlike the *Atlas of Finland*, did not go through subsequent revisions. In fact, only in recent times, about one hundred years after Walker's work, has a general United States national atlas again been issued.

Attempts were made to standardize map symbols in the nineteenth century, notably by von Hauslaub and Peuker in Austria, as noted earlier, and by others including Hassler and the Englishman William Tatham (1752–1819) when they were working in the United States.[37] To the extent that it has been accomplished at all, standardization is a product of twentieth-century international cooperation. Although rampant nationalism in Europe did not end with the nineteenth century, cartography "the science of princes" in the preceding centuries, was to be democratized as well as internationalized in the twentieth century.[38]

# EIGHT

## Modern Cartography:
## Official and Quasi-Official Maps

Modern cartography in this work is divided for convenience into "Official and Quasi-official Maps" (chap. 8) on the one hand and "Private and Institutional Maps" (chap. 9) on the other. As we shall see, there is a good deal of overlap in sources of support, whether governmental or private, as well as some similarity in the resulting product. As before, and as indicated by the titles of these two chapters, the emphasis will be on the map rather than on mapping processes. The latter have undergone remarkable developments in the past hundred years, and many treatises have been devoted to these technical topics. However, as in the preceding chapters, our purpose is to attempt to put the map, which reflects these developments, into a social and cultural context, in this case that of the post–nineteenth-century world.

As indicated before, manned flight, that fundamental desideratum of humans from time immemorial, was accomplished prior to the year 1900, as was viable photography, but their full impact on cartography has only been realized in the twentieth century. Photography grew out of studies on the nature of light, which over the centuries had attracted such workers as Aristotle (384–332 B.C.), Abū ʿAlī al-Hasan (Alhazen) (ca. 965–1039), and Levi ben Gerson (d. 1344), as well as those concerned with the camera obscura: da Vinci, Kircher, Robert Boyle (1627–91), Robert Hooke (1635–1703), and William W. Wallaston (1766–1828), among many others. But it was only after the Frenchmen Joseph-Nicéphore Niepce (1765–1833) and Louis-Jacques-Mandé Daguerre (1789–1851), in collaboration, were able to fix an exposed image, making it resistant to further light action (formally announced in 1839), that true photography was possible. At about the same time, the Englishman William Henry Fox Talbot (1800–1877) developed a method of making many copies from a single negative (1838), and thus modern photography was born.[1]

An aerial platform from which to take photographs was available before the invention of photography in the form of the hot air balloon of the Montgolfier brothers, Jacques-Étienne (1745–99) and Joseph-Michel (1740–1810), of France. At first the passengers were animals and birds, but soon humans were lofted, and shortly the hydrogen-filled balloon was developed. Anchored observational balloons were used by Napoléon Bonaparte (1769–1821) and by both sides during the American Civil War (1861–65). The first aerial photograph was taken in 1838 by the Frenchman Gaspard-Félix Tournachon (1820–1910), who preferred the name Nadar. Nadar made a simple (black-and-white) positive on glass from a tethered balloon 262 feet up, to be followed two years later by one of Boston from an elevation of 1,200 feet by the Americans Samuel A. King (1828–1914) and his assistant, James Wallace Black. Meanwhile, great improvements were being made in optical equipment, especially in Germany, including the binocular stereoscope. What was now needed was a more controlled means of flight, made possible by the heavier-than-air craft (airplane) pioneered in the first decade of the twentieth century by the Wright brothers, Orville (1871–1948) and Wilbur (1867–1912), and others. Photographs were taken from reconnaissance planes during World War I (1914–18), but soon cameras were mounted on actual aircraft and overlapping vertical photos taken at regular intervals. As a result of this development and of the increased altitude soon attained, a new branch of surveying known as photogrammetry was developed in the late 1920s and 1930s. This will be discussed later, but it should be mentioned that only rarely had artists drawn landscapes (as opposed to maps) from a purely vertical view, a notable exception being the "Topography of Niagara" in 1827 by the American painter George Catlin (1796–1872).[2]

In a sense the modern period of cartography can be said to begin with a formal proposal in 1891 for an International Map of the World (IMW) at the scale of 1:1,000,000 (one millimeter on the map represents one kilometer on the ground). The Representative Fraction, or RF, of which "1:1,000,000" is an example, is the most general expression of scale, in contrast to the verbal scale (a description such as that bracketed above) and the graphical scale (in which a line of suitable length is divided and numbered; see for example, fig. 8.7). The RF arose from the metric system, which had been instituted in France in the 1790s as part of the republican reforms of weights and measures. A unit of length, the meter, was defined initially as 1/10,000,000 part of the quadrant of the circumference of the earth from the North Pole through Paris to the equator, and it was thus a product of the rationalism of the French and their experience in topographic surveying and geodesy. The meter was standardized with a platinum, and later a platinum-iridium alloy, bar (eventually expressed in wavelengths of the krypton-86 atom) as 1.093613 yards. This measure was slow to be adopted even in France, but the IMW helped internationalize the metric system, which has still by no means been universally accepted. The *Système international d'unités,* or SI, is the most recent and widely used version of the metric system.

The proposal for the IMW was made by Albrecht Penck (1858– 1945), a German geomorphologist at the University of Vienna, during the Fifth International Geographical Congress held in Berne, Switzerland. Some years before this, Penck had worked out preliminary specifications for the projected map in the form of a series of standard sheets bounded by parallels and meridians and utilizing uniform symbols. Considerable opposition to this idea had to be overcome, but at subsequent congresses the concept gained increasing acceptance. Experimental sheets were produced in several European countries, and in 1909 the First International Map Committee was convened in London to consider general specifications and production methods. Agreement on these was reached at a second meeting of the committee held in Paris in 1913, which also set up a central office at Southampton, the home of Britain's Ordnance Survey. Understandably, little work was accomplished during World War I (1914–18), but after 1921 the central office published reports annually. Considerable progress was made in the production of the map sheets between the world wars, but at the end of this period great areas still lacked coverage, including much of interior and eastern Asia and North America. In the spirit of isolationism after World War I and the decision not to join the American-inspired League of Nations (which after its establishment was a sponsor of the IMW), the United States did not participate in the project, but it did produce a few experimental sheets. In contrast, the Survey of India, a government department, was one of the first to complete its coverage of a large area termed "Greater India." A remarkable contribution to the series was the "Map of Hispanic America on the Millionth Scale," consisting of over one hundred sheets (or about one-tenth of the total in the series) covering South and Central America, by the (private) American Geographical Society of New York.

Although some IMW sheets were published during World War II (1939–45) because of the great increase in flying activity at this time, the need for aeronautical charts became of paramount and immediate concern. To meet this need, a new international map of the world on the scale of 1:1,000,000, the World Aeronautical Chart (WAC), was compiled, generally according to a form drawn up by the International Civil

Aviation Organization. Most of the work was accomplished in the United States under the direction of the U.S. Aeronautical Chart Service, later the Aeronautical Chart and Information Center and now the Defense Mapping Agency (DMA) Aerospace Center. After the Second World War, many of the projects formerly supervised by the League of Nations were taken over by the United Nations, founded in 1945. Thus, in 1953, the functions of the IMW Central Office were transferred to the Department of Economic and Social Affairs (Cartographic Section) of the United Nations in New York. Since this time, a major concern has been the possible duplication of the function of the IMW by the WAC. However, it has now been decided to keep the two world map series separate because their objectives are different, the IMW being for general and scientific purposes and the WAC sheets specifically for the use of the aviator. At the same time, it was agreed that source materials should be available for the compilation of both series and that consideration should be given to use of a common projection and uniform sheet limits.[3]

Before discussing the nature of the cartography of these two most important intermediate-scale international world maps, it is necessary to understand the political climate and technological progress that made these projects possible. We have noticed the secrecy that often surrounded mapmaking activities in the past. On the other hand, we have also alluded to a "brain drain" of scientists (including cartographers) from one country to another and, later, the connecting of triangulation nets across political boundaries. In spite of the international character of these and other operations, from the fifteenth century to the end of the nineteenth, mapping was largely for the benefit of individual sovereign states. It is a tragic irony that at a time when Europe was emerging from its nationalistic parochialism, the continent should be devastated by two world wars. Nevertheless, through all the vicissitudes of the twentieth century, the IMW has been a vehicle for international cooperation, if not an unqualified cartographic success.

The technology that produced the IMW and, later, the WAC projects includes the progress of topographic mapping, the advance of geodesy, and the development of aviation. Sheets of both international series are compiled with information from more detailed sources, especially topographic maps. Topographic mapping, in turn, depends on the accuracy of the triangulation network and the spheroid to which it is related. The Clarke Spheroid of 1866, a theoretical figure, was largely superseded by the International (Hayford) Spheroid adopted by the International Geodetic Association in 1924. Global Positioning Systems, or GPS (discussed in the next chapter), have recently revolutionized geodesy. Pre-

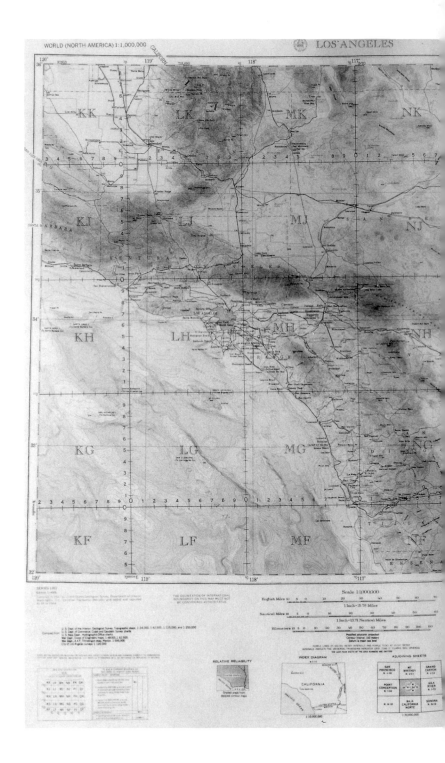

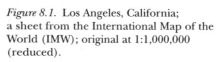

*Figure 8.1.* Los Angeles, California;
a sheet from the International Map of the
World (IMW); original at 1:1,000,000
(reduced).

viously, geodesic work relied on land surveys made by triangulation. With the use of land-based receivers utilizing signals from satellites, more precise measurements of the shape and size of the earth have resulted. Over the centuries we have progressed from a view of the earth as a sphere, through a prolate spheriod, to an oblate spheroid, to a more complicated form of the latter figure. Moreover, GPS has allowed us to measure minor changes in the shape, and therefore the dimensions, of the earth over short periods of time. For maps of small scale, such deviations from a regular geometrical figure are not important, but for those of the largest scales they need to be taken into account. Triangulation networks have been connected between continents, sometimes spanning great distances over water bodies. For example, the Scandinavian system of triangulation has been connected via the Faroe Islands, Iceland, Greenland, and Newfoundland to mainland Canada. This was accomplished by the Hiran method, whereby a measurement is made of the time taken for an electromagnetic impulse to travel from one point to another. Thus the Old World and New World triangulation nets have been connected. We shall consider the impact of the airplane on topographic mapping, as well as the importance of satellites in cartography, in later sections of this chapter.

Figure 8.1 shows an IMW sheet greatly reduced in size. Like all maps in the series it has a unique designation, in this case "N.I 11, Los Angeles." It covers 4 degrees of latitude and 6 degrees of longitude, as do all of the other IMW sheets except those poleward of 60 degrees latitude, which, because of the rapid convergence of meridians, cover 12 degrees longitude or more. The projection used is the polyconic, modified so that adjacent sheets fit on all four sides. The modified polyconic, like its parent projection, is neither conformal (orthomorphic) nor equivalent (equal-area) but is suitable to the purpose and scale of the map. The hyposometric system of relief is used, with principal contours at 200, 500, 1,000, 1,500, 2,000, 2,500 and further increments of 1,000 meters above sea level; the tints between contours progress from green in the lowlands through ochre to brown for the mountains. The entire system works well on the Los Angeles sheet, on which the option to add auxiliary contours every 100 meters has been exercised to show relief on the land from below sea level to 3,500 meters. It is not as satisfactory in regions with lower relief differences and less varied landforms; in fact, a major problem of the IMW series has been the difficulty in creating symbols that are equally applicable to all parts of the world. Undersea features are delineated by isobaths whose numerical values are the same as those used for dry-land contours; the datum for these is mean sea level, and tints of blue are used between selected isobaths, progressively deeper in hue to indicate deeper water. Lettering, boundaries, transportation features, and so on conform to the specifications contained in the Resolutions of the IMW of 1909 and 1913.[4]

We have mentioned that from the outset there was criticism of the IMW; opposition to the series has again developed in recent years. Some critics charge that the scale is not really useful for the intended purposes, and others think that the symbolization, which reflects the technology of the early years of this century, is badly in need of revision (fig. 8.2).[5] When the IMW was initiated, it was envisaged as a general planning map that would serve as a base for maps of other distributions—population, ethnic groups, archaeology, vegetation, soils, and geology. While maps of all of these phenomena have been produced on the IMW base, they are generally of quite limited extent. The most ambitious effort in this direction is a series of geological maps of the former Soviet Union on the scale of 1:1,000,000. To some extent, the functions of the IMW in the former Soviet Union and in Eastern Europe are satisfied by the *Karta Mira* (World Map), on the scale of 1:2,500,000, in 244 sheets. This general world map was prepared by the former Soviet Union and by politically affiliated states, each with special areas of responsibility. Fortunately, this excellent coverage is widely available.

The IMW is still unfinished, even though new sheets are being compiled, and revised sheets of some areas previously mapped in this series have been produced from time to time. By contrast, the WAC, on the scale of 1:1,000,000, is complete in that sheets have been made of all areas of the earth.[6] Superficially, the IMW and WAC sheets are similar in appearance, but there are significant differences between the two series. The Lambert conic conformal projection, used for the WAC, has small scale error and relatively straight azimuths over the extent of a single chart, and, of course, shapes are correct around a point. Like the IMW, WAC sheets cover 4 degrees of latitude; at the equator, both series cover 6 degrees of longitude, with progressively greater longitudinal coverage on sheets nearer to the poles, where the meridians converge rapidly. During World War II, some sheets were printed on cloth to render them less susceptible to moisture damage in case, for example, aircraft personnel had to parachute into bodies of water. A wide variety of information of value to the aerial navigator is contained on the WAC series, which has hypsometric relief broadly similar to that of the IMW. The international Operational Navigation Chart (ONC) with shaded relief now supplements the WAC series. On all of these charts, the quality of the base information varies enormously from area to area depending on the nature of the source materials available for the compilation. Aeronautical charts are produced with smaller scales, such as the 1:5,000,000

## INTERNATIONAL MAP OF THE WORLD
## SCALE 1:1.000.000
## PROPOSED CHANGES TO CONVENTIONAL SIGNS

PROPOSED SYMBOLS          PRESENT IMW SYMBOLS

**RAILWAYS**

Under construction                    Under construction

Two or more tracks (with station) ...........

Single track ........................

Narrow gauge or light ..................

**ROADS**

Dual highway ........................

Main road ........................

Secondary road ........................

Track or path ........................

} RED

**AERODROMES**

Military or civil ........................          With hangar

"   "   "   without facilities ...........          Landing ground

**RIVERS. STREAMS ETC**

Perennial ........................

Sometimes dry ........................

Unsurveyed ........................

Canal navigable ........................

Canal non-navigable ........................

Limit of pack-ice ........................          March

} BLUE

Global Navigation and Planning Chart (GNC) and the 1:2,000,000 Jet Navigation Chart (JNC) for route planning, as well as with larger scales, such as the 1:500,000 Tactical Pilotage Chart (TPC) and the 1:250,000 Joint Operational Graphic (JOG) Air. In addition, approach and proce-

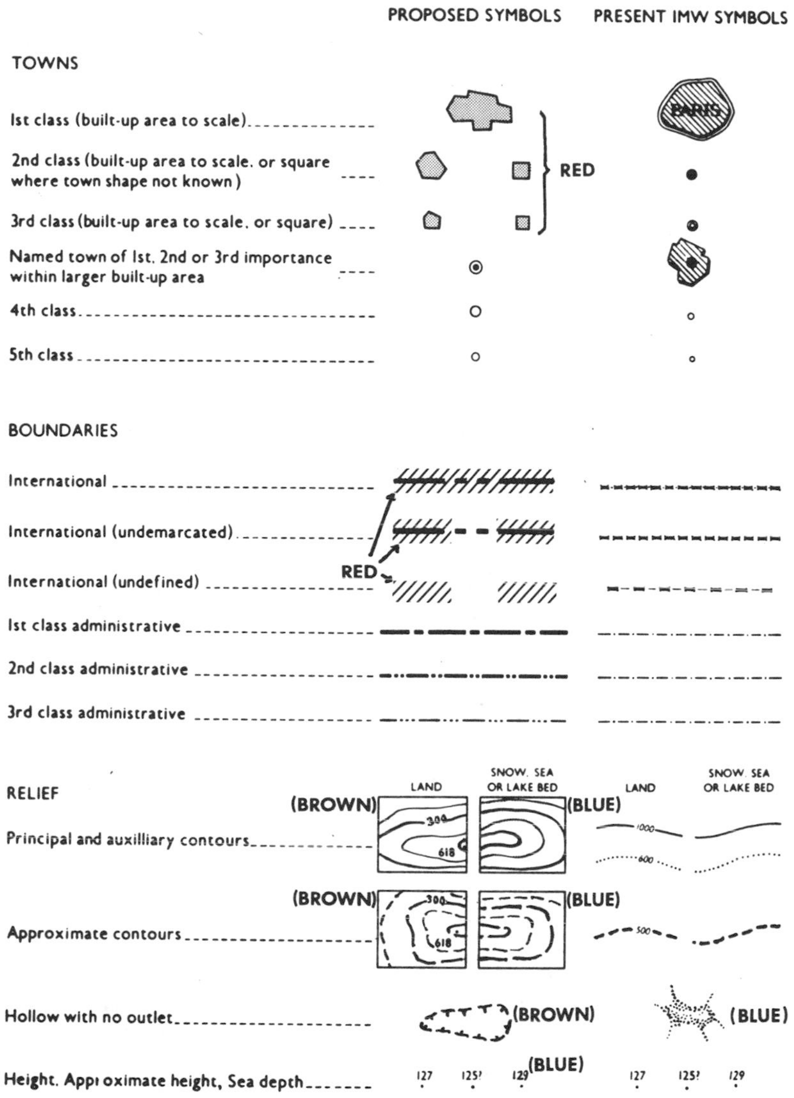

*Figure 8.2.* Existing and proposed symbols for the International Map of the World (slightly reduced)

dure charts are available for specific areas; these now use shaded relief for landforms and are printed in colors suitable to the illumination in the cockpit. Thus the airplane has internationalized cartography in very important and positive ways.

That age-old device the hydrographic chart has recently undergone important changes as a result of the availability of more and better data, mostly since World War II. For example, echo sounding (through the use of automatic depth-finding mechanisms) and new instruments and techniques for determining geographical position at sea are providing abundant information. Echo sounding, or sonar, permits continuous depth traces across the ocean floor to be made by a ship in progress so that, at last, the true three-dimensional form of the oceans can be charted. Symbolization has been improved through the use of color and modern printing techniques. The British Admiralty and the U.S. DMA Hydrographic/Topographic Center, in particular, have produced charts of all the world's oceans, at many scales. An attempt to coordinate the activities of the various hydrographic survey organizations, including charting, has been made through the International Hydrographic Bureau, with headquarters in Monaco.[7] This institution was founded under the patronage of Prince Albert I of Monaco, who in 1903 established the series called the General Bathymetric Chart (GEBCO). These charts are on the Mercator projection with a scale at the equator of 1:10,000,000, while the Arctic and Antarctic regions are on a polar stereographic projection with a scale of 1:6,000,000 at 75 degrees north and south, respectively. Terrain is represented with hypsometric tints on the land (brown) and water (blue), with isobaths in meters.

In addition, the bureau supervises publication of the International (INT) chart series covering all of the oceans at various scales. This ongoing activity has revealed that the morphology of the ocean basins is as varied as that of the land areas, with continental shelves and scarps, great canyons, profound deeps, high mountain ranges, individual seamounts, and so on, in opposition to an earlier idea that the seabeds were rather uniform. The most important discovery from this activity in terms of earth science theory is the presence of midocean ridges on the floors of all the major oceans, from which it is inferred that the present continents have spread predominantly laterally from an original, single (or two) landmass(es). On the basis of the configuration of coasts, as shown on charts, it was early observed that the continental coasts seemed to fit like pieces in a jigsaw puzzle, an idea formalized by the German earth scientist Alfred Wegener (1880–1930). At the time of Wegener's death, not enough evidence was forthcoming to support a viable mechanism to explain "continental displacement," or "continental drift," as it was later termed. But the modified theory of plate tectonics has received almost universal acceptance in recent years from mineralogical, fossil, and morphological as well as cartographic evidence, and it has become the fundamental basis of modern geophysics. In view of these developments, it has been said that most of the earth has been "discovered" in the past half-century. As the resources of the oceans—or "inner space," as they have been called—are increasingly utilized, even better maps of this greater part of the earth's surface will be needed and produced.

In earlier chapters we have traced the development of the large-scale topographic map. We have also alluded to the use of this general type of map as a base for other cartographic works of large scale and as the source for compiled maps of intermediate and small scale. The methods by which topographic maps are made have been revolutionized in this century through the airplane and the development of aerial photography. As indicated earlier, this has given rise to photogrammetry, which is the science of obtaining reliable measurements by means of photography and, by extension, of mapping from photos.

In keeping with our intention to emphasize maps rather than mapping, we will confine our remarks concerning photogrammetry to a bare minimum. However, we should re-emphasize that the use of aerial photographs beginning in the first half of the twentieth century has wrought changes in cartography comparable perhaps only to the effect of printing in the Renaissance or the use of satellites and the computer in the second half of the twentieth century. Photogrammetric methods have remarkably reduced both the cost and the time involved in making maps, made it possible to map areas that would otherwise be difficult to reach, and, most important, increased the quality and accuracy of mapping generally. It is necessary only to compare topographic quadrangles of the same area surveyed by ground and aerial methods to appreciate this point (fig. 8.3).

The preferred source materials for air surveys are vertical photographs with a sixty percent overlap along the line of flight of the aircraft taking pictures (endlap) and a twenty-five percent overlap with adjacent flight lines (sidelap). The endlap and sidelap ensure that only the most accurate part of any particular photo—the center—need be used, and, more significant, make possible stereoscopic analysis. This is accomplished with the aid of the stereoscope and much more elaborate and refined optical instruments (such as the Multiplex) based on the same principle, enabling a viewer to observe a miniature, three-dimensional model of a given area. Such models, made by fusing photographic images of the area taken from different places (stations), utilizing the parallax factor, are used for contouring as well as for planimetric mapping.[8] Because by this method the stereoscopic model, with all its rich detail, is in the photogrammetrist's view, it is obvious why aerial mapping, especially in areas of high relief, is generally superior to field surveys, in which one must interpolate between a finite number of points of known

AERIAL PHOTOGRAPH USED IN THE PREPARATION OF MAP SHOWN BELOW

A PORTION OF THE DELANO, PA., 7.5' QUADRANGLE MAP
Scale 1:24,000.   Contour interval 20 feet.   Mapped in 1946.

A PORTION OF THE MAHANOY, PA., 15' QUADRANGLE MAP
Scale 1:62,500.   Contour interval 20 feet.   Surveyed in 1889.

These maps and the photograph cover the same ground area.   A comparison of the two maps shows the extensive changes that have taken place since the Mahanoy quadrangle was mapped in 1889.   They also illustrate the value of 1:24,000-scale mapping where culture is dense or where greater detail is needed.   Older maps, such as the Mahanoy 1:62,500, are being replaced with modern maps as rapidly as the program permits.

BENCH MARK TABLET

MULTIPLEX

PLANETABLE

A SURVEY MARKER AND SOME OF THE INSTRUMENTS USED TO PREPARE A TOPOGRAPHIC MAP

*Figure 8.3.* Layout showing the relationship between an aerial photo
and a map, as well as maps of different scale and date.
Prepared and published by the United States Geological Survey.

value. Of course, some horizontal and vertical control points to which the aerial photo coverage is tied must be determined in the field; first-, second-, and third-order points (the numbers refer to the degree of accuracy of these control points, which varies in different surveys) used for this purpose are indicated on the ground by metal tablets. After a map is compiled by photogrammetric means and before it is finally rendered, field checking is most desirable, but in the case of areas difficult of access, this is not always undertaken or absolutely necessary.

It might properly be asked at this point why vertical aerial photos are not used in place of topographic maps. The answer is simply that even though distortions can be removed to make them more or less correct in scale, such photos possess too much information; ideally a map represents a judicious selection of data for particular purposes. Actually, annotated mosaics of aerial photos are used to show the texture or character of areas; in some series, the aerial photo coverage is printed on the reverse side of a topographic sheet as a mosaic (uncontrolled) at approximately the same scale as the map. In addition, rectified, controlled photo mosaics are sometimes used as maps, with annotations overprinted in true or false color, as in the pictomap of the DMA Topographic Center (fig. 8.4), which is similar to orthophotomaps of the United States Geological Survey (USGS). Such presentations are particularly valuable when the nature of the surface cover (vegetation, swamps, and so forth) is important to the user. Moreover, orthophotomaps are less costly, speedier to produce, and more easily revised than topographic maps, the subject of the following paragraphs.[9]

In the United States, topographic maps on the 1:62,500 scale (one inch on the map represents approximately one mile on the earth) are now superseded by those on the larger scale of 1:24,000, which will be discussed and illustrated subsequently. Indeed 1:62,500 USGS maps are no longer printed or distributed by the U.S. Government Printing Office (GPO). However, because in many parts of the world maps of similar scale (for example, 1:62,360 or 1:50,000) are more common, and because maps of such scales are of historical importance, we will begin our discussion of twentieth-century topographic maps using as an exemplar of topographic sheets of this general scale range the Orbisonia (Pennsylvania) quadrangle of the USGS 1:62,500 series. Figure 8.5 shows this sheet in contoured form, and figure 8.6 is a contoured and shaded-relief version of the same map for comparison. It is unnecessary to go into great detail concerning symbolization on topographic maps, but in general water features are rendered in blue, cultural features in black or red, and relief in brown. Sometimes broad classes of vegetation are

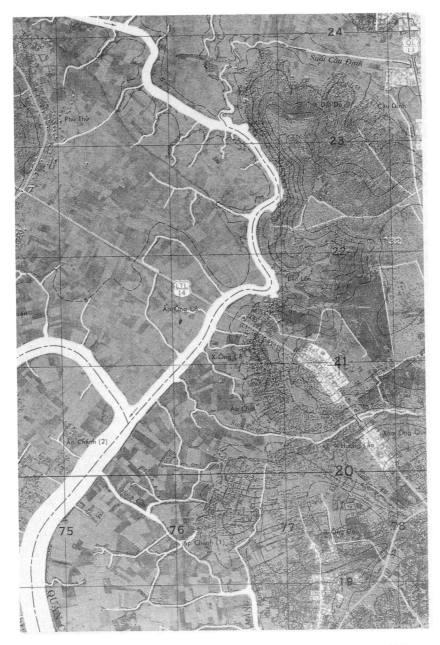

*Figure 8.4.* Portion of pictomap of part of Vietnam (original scale: 1:25,000).
Prepared and published by the Army Map Service, later TOPOCOM and now the
Defense Mapping Agency (DMA) Topographic Center.

added in green. A 1:62,500 USGS quadrangle typically covers 15′ (fif-teen minutes) of latitude by 15′ of longitude (sixty minutes equals 1 degree) and has a contour interval of twenty feet. Other sheet margins and contour intervals are used for particular situations. For example, in areas of very low relief, a five-foot contour interval may be used, and some special sheets, such as those of National Parks, may cover more than 15′ by 15′ to accommodate the whole of such an area on one map.

In the printing of colored maps, especially topographic maps, pho-tolithography is now the usual method of reproduction. The cartogra-pher prepares plates, one for each color, from which negatives can be made. After World War II, a method known as plastic scribing was devel-oped for this purpose. In this process, the cartographer uses a dimen-sionally stable plastic sheet that has a photographically opaque coating applied to it. The operator scribes—that is, scratches through the coat-ing—the desired information and in this way prepares what amounts to a negative, so that the photographic negative stage can be bypassed; obviously, positive images can also be produced from the negatives. From the scribed sheets (one for each color), emulsion-coated metal printing plates, which have replaced those of stone formerly used, can be prepared directly. These flexible metal plates are then introduced into a rotary press, which allows one color after another to be printed with great speed. Plastic scribing has returned control of the production of maps more particularly from the photographer to the cartographer. However, nearly all of these processes have now been computerized, so that cartography has gone from the stone (lithographic) age to the elec-tronic (computer) age in two generations.

Fundamental to the modern topographic map is the contour line, whose ancestry we have discussed earlier. Obviously, however small the contour interval (representation of vertical distance between successive contour lines), important details that may give real character to an area and are observable on aerial photos can be lost between the contours. One may imagine, for example, what would be lost in the representation of a person with a five-foot contour interval. On the other hand, small but significant features can be rendered in the continuous tone of the plastic shading. Interestingly, the landscape shown in figure 8.6 is shaded as though illuminated from the northwest, a direction from which the area never actually receives its light. This convention has a practical basis, since illumination from the bottom of the map produces a pseudoscopic (inverted-image) effect for many viewers. Shading can be produced by hand or with an airbrush, but increasingly, as in so many mapping operations, it is now being accomplished more objectively and more rapidly by means of the computer. The USGS has eschewed using

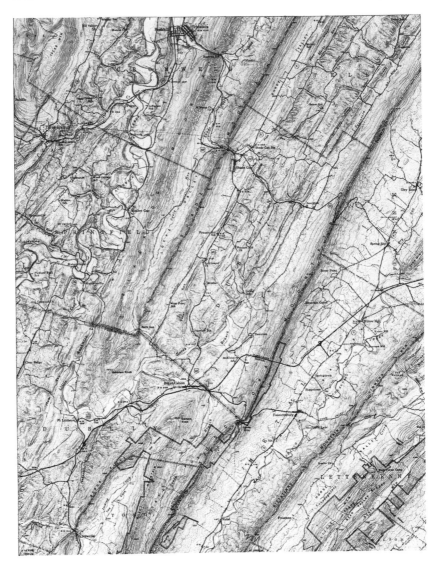

*Figure 8.5.* Contoured topographic map of Orbisonia, Pennsylvania,
by the United States Geological Survey, with marginal data
eliminated (original scale: 1:62,500 reduced).

any kind of subjective lines for showing rocky relief features on its topographic maps. However, on the topographic maps of other countries,
craggy features are often represented by terrain drawing because it is
thought that contours give too rounded an appearance to such landforms as arêtes, cirques, and so on.[10]

*Figure 8.6.* Shaded-relief edition of contour map of the area shown in fig. 8.5
(original scale: 1:62,500) by the United States Geological Survey,
with marginal data eliminated (reduced).

The need of maps for mining, engineering, urban areas and so forth
with greater detail than those of traditional scales has led the primary
mapping agencies of many countries to introduce new series. Thus the
USGS 1:24,000 (7½′ by 7½′) sheets now cover the country. Figure 8.7
is a small part of a map in this series. Many of the remarks we have made

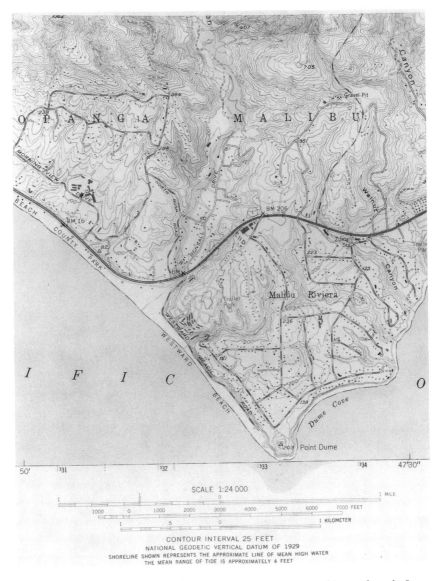

*Figure 8.7.* Detail of the Malibu, California, 1:24,000 topographic quadrangle from the United States Geological Survey with some marginal data below the neat line.

above concerning topographic maps—color, symbolization, contouring, and so on—apply to this series. Of course, such very detailed maps are not suitable for all purposes, so some of the survey departments of the world have prepared maps of their countries at more general scales for regional planning; for example, the USGS publishes topographic maps

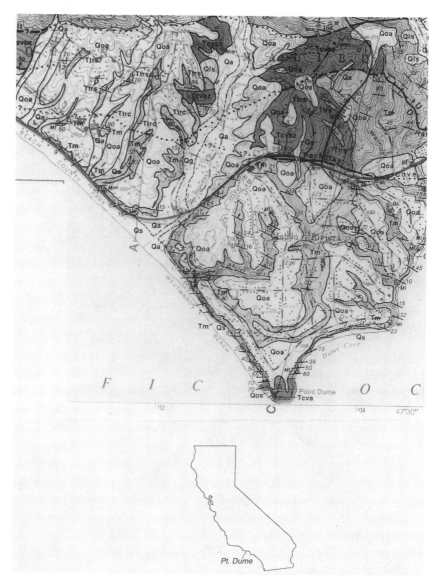

*Figure 8.8.* Detail of a geologic map of the area shown in fig. 8.7.

of 1:100,000 and 1:250,000 scales. There is a desire on the part of the USGS to metricate, and on a limited 7½′ by 15′ series the contours and elevations (spot heights) are shown in meters. A cause of some difficulty is that the USGS is not the sole authority for domestic topographic mapping in the United States in the sense that the Ordnance Survey is in

Britain; the duties of topographic mapping in the United States are shared between a number of government agencies. Another problem is that topographic mapping is often initiated by the states and financed on a shared-cost basis.

The United States Geological Survey was founded in 1879 as a result of the remarkable mapping activities associated with the detailed exploration and opening up of the arid and semiarid American West, as mentioned earlier. Its purpose was making "a systematic study of the geology and natural resources of the United States and . . . classifying the public lands."[11] The foundation of the USGS thus differed from the older topographic surveys in Europe, which were more strictly geographical, topographical, or military in outlook. Having been brought into being after the science of geology was well developed, and in accordance with the terms of its establishment and title, the USGS in its early years paid particular attention to the physical landscape. For example, the relief rendering on early USGS maps is among the best in the world for the time; less importance was attached to cultural features, in comparison with the maps of a number of other surveys. However, because of the increasing importance of human landmarks upon the earth, USGS maps, particularly those of the 1:24,000 series, are now much richer in cultural detail. The USGS maintains a national Earth Science Information Center (ESIC) in Reston, Virginia, near Washington, D.C., to provide data on maps, aerial photos, and geodetic control in the United States and its territories and possessions as well as a dozen regional ESIC offices and other offices for mineral and water information.

Of course not all of the land surface of the earth has been covered by up-to-date topographic maps suitable for engineering purposes.[12] Generally, European countries are well mapped topographically. France and Britain were surveyed uniformly but, because of political fragmentation until the middle of the last century, Germany and Italy did not have centralized control of surveys when they were initiated. This later caused problems in the mapping of those countries that have now been generally overcome. Some former colonial areas are often surprisingly well covered by topographic maps; for example, India is much better mapped today than China. After World War II, Britain and France established agencies to assist certain developing countries in their topographic mapping endeavors, and a number of private companies in the West have engaged in similar activities. It can be said, with truth, that we do not really know an area (its resources, morphology, size, and so on) until it has been mapped in detail.[13] Furthermore, many human activities, including the compilation of other kinds of maps, cannot be accomplished without good topographic coverage.

We have mentioned geological and land-use maps as two classes that are fundamentally dependent on the topographic base. Detailed geological mapping is generally performed on a piecemeal basis according to need. A geological map is one that represents the earth with the overburden (soil or other friable material) stripped off to reveal the rocky crust. On such maps, the age of the rocks, as well as their lithologic type, is represented by a series of conventional colors and symbols (fig. 8.8).[14] These are among the most complicated maps, but simplified and generalized geological maps have been made, as indicated previously, of continental, subcontinental, and national areas. In addition, lithological maps—on which the type but not the age of the rock is indicated—as well as soil maps have been produced for a variety of economic purposes (quarrying, agriculture, and so on). Geological maps are frequently accompanied by selected cross sections or profiles showing structural relationships today, as in the earliest examples (see figs. 7.1 and 7.2 in chap. 7).

Land-use or land-cover maps have a distinguished lineage, as demonstrated earlier, and, for ecological reasons related to increasing pressure on the land, they are of great contemporary interest. One of the most remarkable national mapping enterprises in recent times is the Land Utilisation Survey of Great Britain, conducted under the direction of Professor (later Sir) L. Dudley Stamp. The principal aims of this survey, which came into being in 1930, were to "make a record of the existing use of every acre in England, Wales and Scotland" and to "secure the support of a well informed public opinion for the work of planning the land for the future."[15] A number of the most important individuals involved in British economic and environmental planning in the 1930s were connected with the project, but the fieldwork was accomplished largely by secondary-school students under the supervision of their geography teachers. The compilation scale was six inches to one mile; the sheets were reduced to one inch to one mile (1:63,360) and published on the Ordnance Survey base maps of this scale (fig. 8.9). Land uses are indicated by colors as follows: forest and woodland (dark green); meadow and permanent grass (light green); arable land (brown); heathland (yellow); gardens and orchards (purple); buildings, yards, roads, and so on (red). This color scheme has now become more or less conventional for land-use maps, just as the topographic map color scheme indicated earlier is conventional for maps of that type. By 1940 the Land Utilisation Survey of Britain was virtually complete, and it proved of enormous value as Britain systematically expanded its agricultural production during World War II. The British experience in land-use mapping has had several important results: (1) It has provided a base of

# THE LAND UTILISATION SURVEY OF BRITAIN
## Specimen of a " One-Inch " map

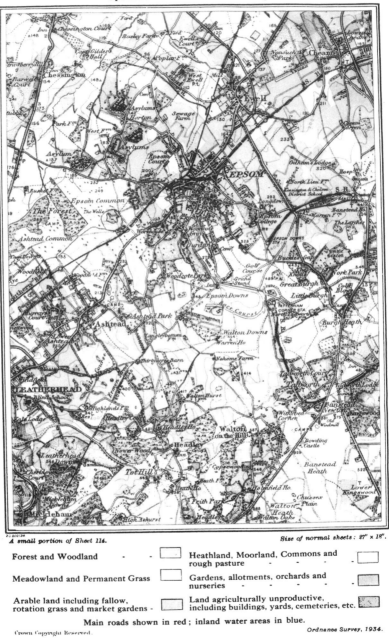

A small portion of Sheet 114.

Size of normal sheets: 27" x 18".

| | | |
|---|---|---|
| Forest and Woodland | Heathland, Moorland, Commons and rough pasture | |
| Meadowland and Permanent Grass | Gardens, allotments, orchards and nurseries | |
| Arable land including fallow, rotation grass and market gardens | Land agriculturally unproductive, including buildings, yards, cemeteries, etc. | |

Main roads shown in red ; inland water areas in blue.

Ordnance Survey, 1934.

*Figure 8.9.* Specimen sheet with legend of the Land Utilisation Survey of Britain (original scale: 1:63,360) (reduced).

comparison with past and future land uses; (2) it has given rise to even more detailed (1:25,000) land-use mapping activities in Britain; and (3) it has led to the formation of a commission of the International Geographical Union (IGU) appointed to apply the principles to other countries, especially developing nations. Land-use mapping of extensive areas has been accomplished almost exclusively with satellite imagery as a data source.[16] Other distributions and subjects that have been mapped using remote sensing include weather and climate, geology, soils, forestry, water resources, wildlife, ecology, archaeology, and urban and regional geography. In order to appreciate how this has come about, it is necessary for us to review, briefly, various space programs.

As with so many developments, rocketry and the associated propellant of gunpowder (explosives) appeared in China and Islam before they were "invented" in the West. After artillery became available in Europe in the later Middle Ages, society was transformed because there was no longer security behind the medieval castle wall. In time greater accuracy was attained in artillery as some of the most fertile minds in Europe studied ballistics, while newly constructed fortresses on shore (for some of which remarkable plans were made) and warships at sea were essentially platforms for guns. However, it was some time before there was power enough to propel a missile skyward any considerable distance and much longer before rockets were used as platforms for imaging the earth. Apparently this first occurred when a camera was fitted to a rocket by German artillery in 1910 (applying theoretical ideas of Swedish industrialist/philanthropist Alfred Nobel), marking the beginning of controlled, unmanned remote sensing.

Phenomenal progress was made through the German experience with V-2 rocketry during World War II. When, in about 1945, German rocket scientists joined the incipient U.S. space program, advanced by Robert Goddard (1882–1945) in Massachusetts and Theodor von Kármán (1881–1963) in California, as well as that of the then–USSR, extraterrestrial flight became possible. From 1960 on, TIROS (a series of meteorological satellites) were launched in the United States and demonstrated a great potential for weather mapping, the first field of cartography to be affected by the new technology. Meanwhile, the Soviets launched Sputnik in 1957, imaged the far side of the moon in 1959, and put a human in space in 1961. The next year marked the first manned space flight by the United States, which soon began a series of missions imaging the earth from space—Gemini (1965–66) and Apollo (1968–69)—with the use of hand-held cameras loaded with color and, later, color infrared (CIR) film, which had been perfected during World War II. The United States Apollo Eleven mission of July 1969 landed two

men on the moon, and many images were taken of this body and of
the earth.

The next development was continuous, extensive surveillance of the
surface of the planet, first accomplished by the Earth Resources Technology Satellite (ERTS) in 1972.[17] Another similar unmanned satellite
was launched in 1975, and the program was renamed Landsat. Since
that time, the surface of the earth (except the polar regions, which the
system does not cover) has been scanned by Landsat as frequently as
every nine days (depending, of course, on cloud cover). By international
agreement, Landsat imagery, which is telemetered to the earth in at least
four multispectral bands, is available to users in any part of the world.[18]
Although not maps themselves, Landsat images have been overlaid or
annotated and used in place of maps (fig. 8.10) and also employed for
map revision. More conventional overlapping vertical aerial photography is not replaced by space imagery, which up to this point has provided
minimal capability for stereoscopy. Both data sources have their own and
complementary uses. The French SPOT and various Russian satellite
programs produce very high quality images, although unlike the Landsat, they do not have universal, continuous satellite coverage of the
earth.

One of the problems with Landsat has been its low spatial resolution—that is, the ability of the system to render a sharply defined image—but this has not prevented its use in monitoring the earth's resources, as intended, or in the creation of mosaics such as the National
Geographic Society's "Space Portrait U.S.A." A great deal of nonsense
has been written about whether such presentations are "reality," but this
begs the question. To a chicken the object called a map is of no value as
a map, but to humans capable of understanding their symbolization,
maps and images convey information of greater or lesser importance
depending on the needs and comprehension of the viewer.[19]

The U.S. National Aeronautics and Space Administration (NASA)
came into being in 1958, and since that time one of the most successful
applications of its missions has been in respect to the atmosphere. Over
the years a number of geostationary and other satellites with various ac-
ronyms—ESSA, ATS, GOES, NOAA—have been launched. In aggregate, these high-altitude satellites have been fitted with a variety of sen-
sors—television, infrared, vidicon, radar—which have given scientists an
immense data source with which to better understand the mechanics of
the atmosphere, to appreciate climate, and to forecast weather conditions. These missions produce images, not strictly maps, but many have
overlays of grids, coastlines, political boundaries, and frequently annota-

tions (both numerical and verbal) that make them maplike. Some of this information filters down to the public through televised weather reports, which are often animated presentations.

Perhaps the most familiar official map to many people is the Daily Weather Map, which is reprinted in simplified form in newspapers. The United States Daily Weather Map originates with the National Oceanic and Atmospheric Administration (NOAA), formerly the Environmental Science Services Administration (ESSA)/Environmental Data Service, and includes the National Weather Service, formerly the U.S. Weather Bureau, Department of Commerce. NOAA maintains over twelve thousand weather stations, of which about two hundred are classed as first-order stations. These last, located in major cities and at airports, are staffed with professional forecasters who furnish meteorological data in the form of daily reports. Every twenty-four hours the National Meteorological Center near Washington, D.C., receives more than twenty-two thousand hourly surface reports and some eight thousand international six-hourly surface reports, as well as a smaller number of reports from ships, balloons, and other locations. Transmissions are made in abbreviated numerical code by radio or telephone line. These data are then plotted on maps. The Daily Weather Map shown in figure 8.11 was compiled from observations made at 7:00 A.M. Eastern Standard Time on Sunday, 15 January 1995.

In discussing the symbolization on the Daily Weather Map, it is useful to recall that this device, of value to almost everyone in the country, came into being after a long period of gestation. We have already discussed the wind chart developed by Halley, isolines applied to climatological phenomena by Humboldt, and other symbolization by Beaufort. Among the earliest climatological maps on which a variety of phenomena appear together are those of the Americans Elias Loomis in the 1840s and Commodore Maury in the next decade. These maps show a composite of weather over a considerable period of time, differing in this respect from the Daily Weather Map, which is a "snapshot" of weather conditions. Nevertheless, the symbols on contemporary weather maps developed from earlier climatological map symbols—for example, arrows used for wind direction and isolines for barometric pressure (isobars)—according to an international code. The arrow symbol has been refined to show not only wind direction but wind speed as well. This is accomplished by "feathers" on the arrows, the number and length of which indicate the speed of the wind in knots. Isobars (with pressure reduced to sea level) are drawn at intervals of four millibars, and high and low pressure centers are noted. Other line symbols are used to show

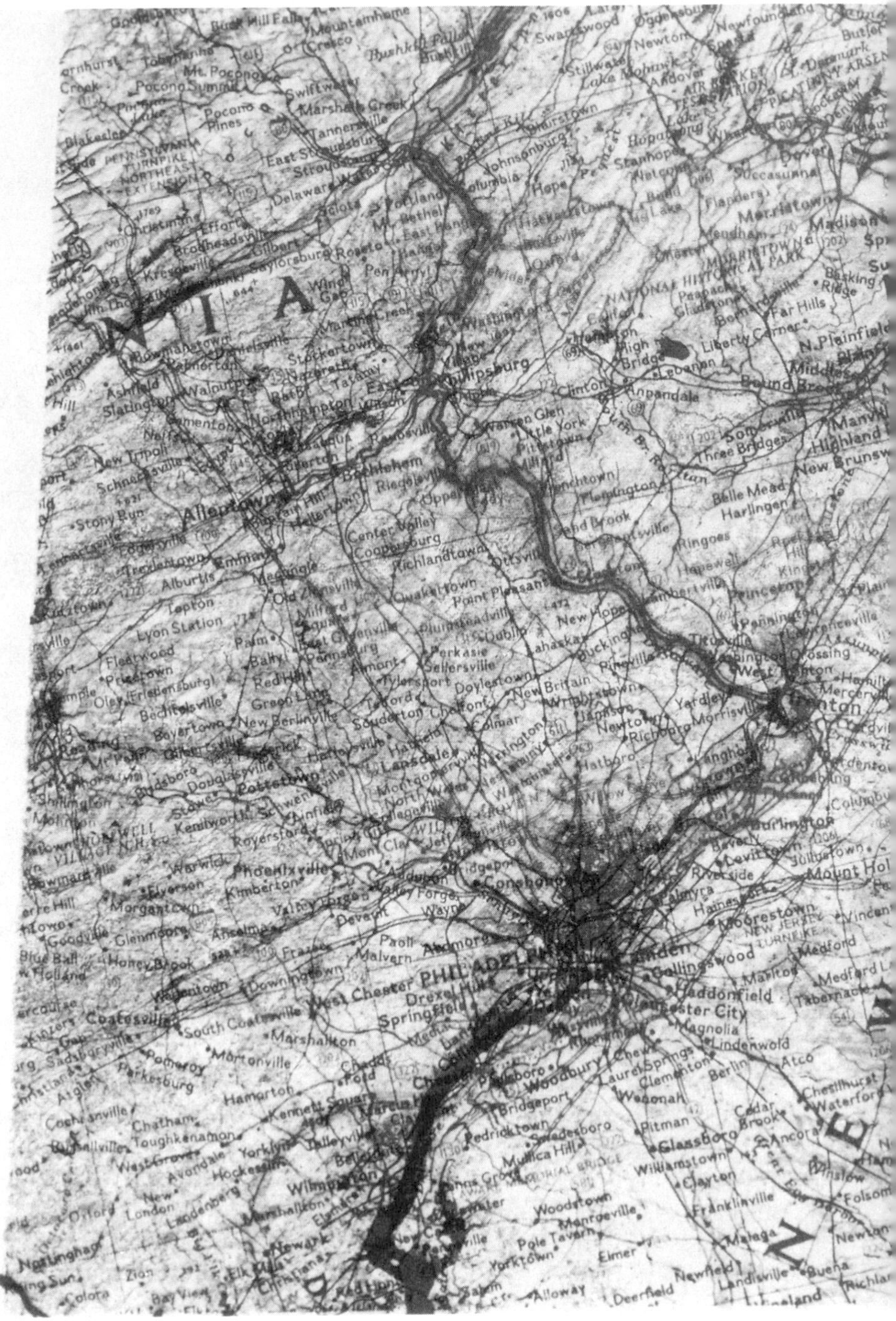
Swartswood
Newton
Stillwater
Andover
Mt. Pocono
Pocono Pines
Swiftwater
Marshalls Creek
Johnsonburg
Stanhope
Flanders
Morristown
Mendham
Madison
Dover
Delaware Water Gap
Stroudsburg
East Stroudsburg
Brodheadsville
Effort
Gilbert
Kresgeville
Hope
Oxford
Pen Argyl
Bangor
Wind Gap
Roseto
NATIONAL HISTORICAL PARK
Peapack
Gladstone
Bernardsville
Far Hills
Basking Ridge
Liberty Corner
N. Plainfield
Bound Brook
High Bridge
Annandale
Clinton
Lebanon
Phillipsburg
Washington
Stockertown
Nazareth
Tatamy
Easton
Ashfield
Slatington
Walnutport
Northampton
New Tripoli
Schnecksville
Stony Run
Allentown
Bethlehem
Riegelsville
Hellertown
Center Valley
Coopersburg
Richlandtown
Quakertown
Frenchtown
Flemington
Little York
Pittstown
Somerville
Three Bridges
Highland Park
Manville
New Brunswick
Belle Mead
Harlingen
Ringoes
Hopewall
Princeton
Pennington
Lawrenceville
Titusville
Trenton
West Trenton
Hamilton
Mercerville
New Hope
Lahaska
Point Pleasant
Plumsteadville
Perkasie
Sellersville
Doylestown
Chalfont
New Britain
Yardley
Morrisville
Newtown
Richboro
Hatboro
Alburtis
Lyon Station
Fleetwood
Oley (Friedensburg)
Bechtelsville
Boyertown
Green Lane
Red Hill
Palm
Bally
Pennsburg
Telford
Souderton
Hatfield
Lansdale
North Wales
Montgomeryville
Pottstown
Douglassville
Kenilworth
Royersford
Phoenixville
Kimberton
Valley Forge
Wayne
Norristown
Conshohocken
Morgantown
Glenmoore
Honey Brook
Downingtown
Paoli
Malvern
Ardmore
West Chester
PHILADELPHIA
Drexel Hill
Springfield
Coatesville
South Coatesville
Marshallton
Mortonville
Pomeroy
Parkesburg
Atglen
Cochranville
Chatham
Toughkenamon
West Grove
Avondale
Kennett Square
Hockessin
Wilmington
New London
Landenberg
Newark
Elkton
Chester
Woodbury
Paulsboro
Bridgeport
Swedesboro
Pedricktown
Mullica Hill
Woodstown
Pole Tavern
Yorktown
Elmer
Alloway
Deerfield
Newfield
Landisville
Buena
Malaga
Franklinville
Clayton
Williamstown
Glassboro
Pitman
Wenonah
Clementon
Berlin
Atco
Lindenwold
Magnolia
Haddonfield
Collingswood
Camden
Moorestown
Marlton
Medford
Mount Holly
Burlington
Levittown
Riverside
Palmyra
Columbus
Bordentown
Waterford
Chesilhurst
Cedar Brook
Ancora
Winslow
Folsom
Newtonville
Langhorne
Mount Holly

*Figure 8.10.* Landsat image (scene) of part of the eastern United States, with map information overlay.

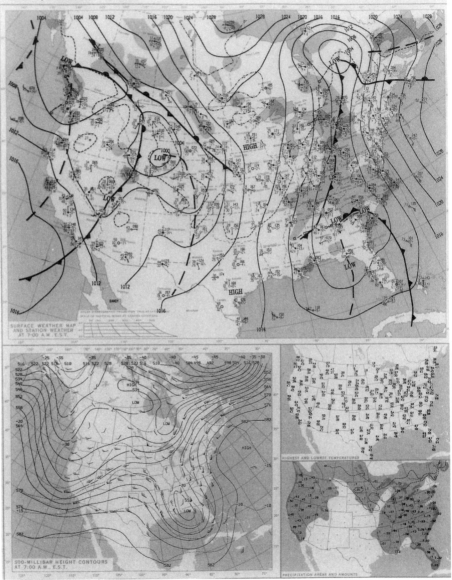

*Figure 8.11.* Summary sheet of Daily Weather Map information.

fronts—whether warm, cold, occluded, or stationary—and troughs of low pressure (dashed lines). Air temperature, dew-point temperature, cloud cover, and visibility data, as well as wind and barometric information, appear in the form of point symbols at selected stations scattered over the map. Areal symbols are used for precipitation, with rain and snow differentiated. Theoretically, some two hundred symbols might appear on the Daily Weather Map, which is a masterpiece of condensation. The completed map is photoelectrically produced near Washington and relayed to stations, which have facsimile machines and computer terminals capable of reproducing the original. In addition to the eight daily synoptic charts (produced at three-hour intervals), the machines process prognostic maps, constant-level and constant-pressure charts of the upper air, and wind-aloft charts.[20]

A sequential series of weather maps can show the passage of pressure cells and fronts and the growth and dissipation of storms. To make this information more graphic, the three-hourly weather analysis maps and six-hourly prognostic maps have been put on film to produce time-lapse movies. These reveal patterns that a more static series of maps might not show and represent a step toward animated cartography, which will be discussed in the next chapter. Satellite images, received hourly (and even more frequently during severe weather events such as hurricanes), are now used in animated form to illustrate and track storm development and movement, with or without annotations. When we realize that such dynamic patterns were first recognized only at the time of the Crimean War (1853–56), we can appreciate that progress in this field has been phenomenal.

Many government departments other than those specifically charged with mapping have important cartographic sections. In the United States, a number of federal agencies produce not only maps to serve their own departmental needs but also thematic maps of interest to the general public. Outstanding among these are small-scale thematic maps of the entire United States by the U.S. Department of Agriculture and the Bureau of the Census. The Department of Agriculture has a long tradition of support of thematic mapping, and several of its employees, such as Oliver E. Baker (1883–1949) and Francis J. Marschner (1882–1966), must be numbered among America's most distinguished cartographers.[21] Over the years, maps of the United States published by the Bureau of the Census at intermediate and small scales have covered a wide variety of topics arising from decennial census returns. General categories include population, housing, and income, with specific maps such as "Metropolitan Statistical Areas" and "Population: Urban and Rural."[22] It is well understood that maps of population, including ethnicity, literacy, and health, are as important for cultural and social studies as physical maps are for mining, agriculture, and engineering. In general, physical maps have received more attention from officialdom than those on human and biological topics, but this is changing as maps from census bureaus and from other government departments in the United States and elsewhere find important use in the classroom as mounted maps or as reference items in sheet map collections.

In the United States, most cadastral and urban mapping now, as in earlier periods, is accomplished at the local level—county, civil township, city, and so forth—and much of the record is in manuscript form. The lack of comparability in the information is particularly critical in today's cities, with their pressing problems. Many years ago a plea was made for the coordination of city surveying and mapping activities in the United States. It was pointed out that in a given city, the interdepartmental and private demands for map information often involve a serious wastage of time because of the lack of a centralized cartographic clearinghouse.[23] The situation has not improved markedly since that time and is now critical, especially in those cases in which one urbanized area abuts another. Political, legal, and financial aspects of municipal surveying and mapping demand more coordination of these activities. The suggestion has been made that, at a minimum, a centralized city survey and mapping agency should be founded to establish standards, to educate both surveyors and the public, and to eliminate wasteful duplication of effort. The problems in Great Britain are much simpler, but in any case in that country large-scale plans of 1:2,500 (approximately twenty-five inches to one mile) are available for the entire kingdom, except some areas of mountain and moorland. In addition, Britain has more detailed plans (1:1,250); such maps include all municipal boundaries, roads, alleys, public buildings (named), houses (numbered), and even smaller structures. To print and keep such maps up to date is a formidable task, but there, as elsewhere, the computer has simplified many procedures in urban and other mapping activities, especially in the matter of revision.

The quintessential official cartographic product is the national atlas, which expresses the pride and independence of the country. The *Atlas of Finland (Suomen Kartasto),* with editions dated 1899, 1910, 1925, and 1993, is usually considered the first true national atlas with a continuous existence, as indicated earlier. It is sometimes difficult to determine whether an atlas of a country, group of states, state, or province is official, quasi-official, or private. An atlas might be sponsored by a government department, compiled at a university, and printed by a private company. In any event, increasing numbers of atlases dealing with discrete political units have appeared in recent years. Some of these focus on a particular set of data within the selected area, but more often these national or state atlases deal with a very wide range of information, as, for example, the *National Atlas of the United States of America* (1970) by the USGS.[24] National atlases have been published by technologically advanced countries such as Canada and France and also by developing ones such as Kenya and Uganda. The maps in the *Atlas of Finland*—covering, as they do, various aspects of geology, climate, vegetation, population, agriculture, forestry, industry, transportation, trade, finance, education, health, and elections as well as political/historical data and geographic regions—may be taken as representative of the contents of an excellent national atlas. Naturally, the subjects selected are those that are important to the particular country represented, and no satisfactory format for all national atlases should or could be devised. There is a great variety in the quality of the maps found in national atlases, depending on the reliability of the census data, the ingenuity of the cartographers, and the skill of the printer. A large number of countries—old and newly founded, rich and poor, large and small—now boast a national atlas.

Many of the above remarks are true of atlases of regional, state, and provincial areas. Again, commercial concerns, universities, and government departments are often jointly involved in the production of such works. Thus the *Atlas of the State of South Carolina* (1825), the first such project in the United States, was authorized by the state legislature, compiled by an individual, and printed commercially.[25] Increasingly, universities, chambers of commerce, and state and provincial governments have concerned themselves with such ventures. Outstanding among these works are the *British Columbia Atlas of Resources* and the *Economic Atlas of Ontario*.[26] These atlases present a wealth of information about the resources of their respective provinces. They should serve as models for states and provinces lacking such coverage or with inadequate atlases.

We should not conclude this chapter on contemporary official cartography before discussing the progress of extraterrestrial mapping. Pioneer work in this field by Galileo and the celestial charting activities of Hevelius and Halley have already been mentioned. These scientists were followed by a succession of astronomers who interested themselves in lunar mapping, an activity that, in its development, can be considered a microcosm of geocartography, albeit with some important differences. Most of the early lunar maps were drawn on the orthographic projection, in which a sphere appears as viewed from an infinite distance. Such is the case of the moon map of Franciscus Fontana (1600–1650), who was also the first person to observe the markings on Mars and to sketch this planet. In this period the usual method of representing the relief of the moon was line shading, although Hevelius had used "molehill" forms on one of his maps.[27] The common system of nomenclature for lunar features—naming them after great scientists and others—was devised by Jean (Giovanni Battista) Riccioli (1598–1671) in the 1650s and to a large extent displaced Hevelius's method of using the names of prominent earth features for the purpose, although some of this terminology is still in use.[28] Like Hevelius, Riccioli showed the moon's librations (oscillations that enable us to see more than half of the lunar surface from the earth) by intersecting circles. J. D. Cassini made lunar maps in the 1680s, but, though artistic, these represented no real scientific advance over the work of Hevelius.

In the eighteenth century, Johann Tobias Mayer Sr. (1723–62) used a micrometer to measure lunar elevations and locations and devised a net of equatorial coordinates for the lunar surface. Lambert (see chap. 6 for his work on projections) employed a conformal projection, the stereographic, for lunar mapping; this allows features, including peripheral ones, to be shown with a minimum of distortion. Of course, a distant view of the subject, with its obvious advantages, has always been available for half of that body to lunar mapmakers, but only recently have cartographers had a similar advantage for terrestrial mapping through aerial and, later, satellite photography. Two of the greatest selenographers of the nineteenth century were Johann H. von Mädler (1794–1874) and Wilhelm Beer, who collaborated on a lunar chart for which they used triangulation. They employed hachures to represent lunar relief, while later in the century Étienne Léopold Trouvelot (1827–95) used a double or triple circle (form-line) method. This was a step toward a contour map of the moon, made possible through lunar photography, which began in the 1850s. Opposition to the use of the contour for selenographical purposes had to be overcome, but this method combined with shaded relief, or with altitude tints between contours, has now become the usual means of representing the surface of the moon.

In recent years, since the earliest manned space flights, and the earlier return of pictures of the unobserved side of the moon from the unmanned *Luna 3* Soviet spacecraft in October 1959, there has been a great flurry of interest in lunar mapping. While previously individual scientists labored for a lifetime to produce a representation, now teams of cartographers from large official mapmaking agencies are engaged in the work. In the United States, for example, the USGS, various other mapping agencies, and the military establishment have been active in lunar mapping since the late 1950s, especially in preparation for the first landing of a human on the moon on 20 July 1969. Without attempting to list all of the lunar maps produced by these agencies, it can be said that they range from single sheets of small scale to multisheet series at 1:1,000,000 (such as the Lunar Aeronautical Charts of the DMA/Aerospace Center) and larger scales. A number of projections other than the orthographic and stereographic are now used for lunar mapping, including the Lambert conformal and Mercator projections. The subjects covered include relief, physiography, tectonics, surface materials, and "geology." Purists have taken exception to this last term being applied to the lunar lithology and have preferred the expression "selenology." More general terms such as "cosmology" and "cosmography" may have to be revived; happily, "cartography," "chart," "plan," "globe," and "map" are general terms equally applicable to terrestrial and extraterrestrial phenomena. As indicated, a variety of techniques has been used to express the surface features of the moon: shaded relief, contours, hypsometric tints, and so on. As in terrestrial mapping, the combination of these methods is facilitated by color printing. A problem in representing the lunar surface by contours has been the lack of a "natural" datum, such as mean sea level on the earth. Frequently the center of the small crater near the lunar equator, Mösting A, is utilized as the vertical datum for contouring, and thus a point, rather than a surface, is used for this purpose.

Celestial bodies other than the moon also command the attention of modern cartographers. A map of Mars at the scale of 1:10,000,000 on the orthographic projection, based on the notes of a number of astronomers, has been prepared by the DMA Topographic Center. The USGS produced an *Atlas of Mars* (consisting largely of photos) at the scale of 1:500,000 and maps of the same planet at 1:5,000,000 in thirty sheets. Detailed maps have also been made of Jupiter (and its satellites), as well as of the planets Venus and Mercury, using remote-sensing techniques first employed for mapping the surface of the earth.[29] This activity is expected to increase as more and better data become available through expanded space programs.

Figure 8.12 is used to illustrate contemporary extraterrestrial mapping. It shows a small part of the surface of the moon, centered on the crater Copernicus, from the DMA Lunar Map on the scale of 1:2,500,000. A combination of contouring and shading is employed for the relief representation. Both regular and depression contours are evident on the larger craters, while small craters are marked by uniform circles. Prominent ridges of less than the contour interval (one thou-

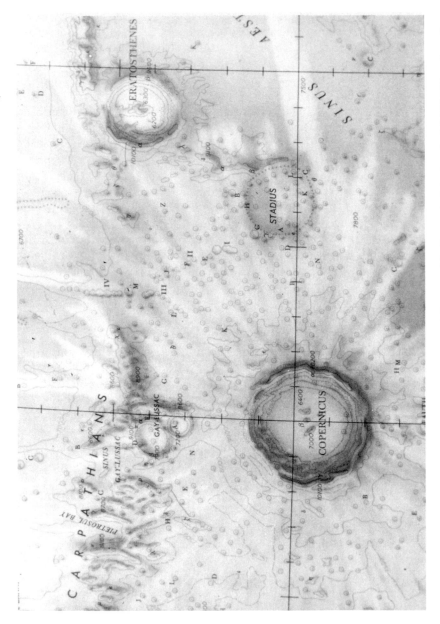

*Figure 8.12.* Small section of a lunar chart from the DMA Topographic Center (original scale: 1:2,500,000; reduced).

sand meters, with supplementary contours at five hundred meters) are marked by a series of dots, as at Stadius. Color is used to distinguish rays (light), craters (yellow to brown), and maria (green to blue). The map is drawn on a modified stereographic projection, related to a selenodetic sphere, with relief data compiled by stereophotogrammetric methods.

We can now return to the earth to conclude our summary of cartography and imaging through the ages with a consideration of contemporary maps predominantly in the private sector.

# NINE

## Modern Cartography:
## Private and Institutional Maps

Although many mapmaking operations, as we have seen, are now largely the province of government agencies, the private sector is extremely important in modern cartography. In general, commercial map companies, geographical societies, and universities, all of which engage in research, tend to dominate nonofficial mapping, although some maps are published by individuals. At a cartographic convention one finds representatives of the business and academic communities and freelance mapmakers as well as workers from government departments. The theme of this book is to show how the map reflects society and culture. Thus in this chapter, which illustrates private and institutional cartography throughout the past hundred years, development is emphasized more than the present state of the art, which in any case will be history tomorrow. Accordingly, maps made in the early and mid–twentieth century, as in the preceding chapter, are included as steps along the way to a later but not necessarily "better" cartography. In any human activity, towering figures appear whose work is imitated but not always improved upon and in some cases may blind later workers to other profitable directions. This has been the case in cartography in the past, as in the examples of Ptolemy and Mercator, and although less true in recent times because of team research and multiple authorship, it is still operative to a degree.[1]

In chapter 7 we discussed a number of geographical "firsts," the quest for which stimulated mapping of certain inaccessible places in the nineteenth century. This continued without interruption into the twentieth century. After many attempts by the sailors of several nations, especially Britain, the Northwest Passage (north of North America) was completely navigated between 1903 and 1906 by the Norwegian Roald Engelbrecht Gravning Amundsen (1872–1928). During the 350 years since Europeans had attempted this geographical "first," the Arctic coasts had been charted in increasingly great detail. Amundsen determined the precise position of the Magnetic North Pole (which had been approximately located by James Ross in 1831) during his traverse. The location of the magnetic poles nonetheless varies through time within certain limits. Later, Amundsen's knowledge of the terrain of Antarctica, gained through fieldwork and, in part, mapping, led to his success in reaching the geographical South Pole on 14 December 1911, a month before his rival, the Englishman Robert Falcon Scott (1868–1912), who perished with his party on their return to base camp. Scott had taken a longer and more difficult route than Amundsen, but through both expeditions much terrain was mapped for the first time. Amundsen had initially contemplated an expedition to the North Pole, but after learning that this geographical first had been attained in April 1909 by the American Robert Edwin Peary (1856–1920), Amundsen turned southward. There is some question as to whether Peary, with the African American Matthew Henson, actually reached the North Pole. However, years of scientific work, including mapping, had preceded this accomplishment in the Arctic, as, indeed, with other geographical firsts in other parts of the globe.

A mid–twentieth-century first was the conquest of Mount Everest on 29 May 1953 by Edmund (later Sir Edmund) Hillary and Sherpa Tensing Norgay. This was the culmination of a long period of preparation—successes in climbing other mountains in the Himalayas, photography (including aerial photography), and photogrammetry and mapping, largely under the aegis of the (private) Royal Geographical Society of London. A map on the scale of 1:50,000 compiled by Arthur R. Hinks (1873–1945), secretary of the R.G.S., and the society's draftsman, H. F. Milne—the so-called Hinks–Milne map—was used to plan the ascent of Everest. Following the successful accomplishment of this, other maps were made of the region, including those by the Federation Français de la Montagne, 1:50,000 (1954–55), and by the American Bradford Washburn, director of the Boston Museum of Science, 1:10,000 (1988). On this last, the delineation of the sharp features of Everest is rendered with a combination of hachures and contours in the manner of the Swiss Alpine maps.[2] Much scientific mapping followed, including that of the International Geophysical Year (IGY) in 1958, focusing on Antarctica.

In our consideration of certain highlights of official cartography in the twentieth century, we noticed that in some instances it is difficult to make a clear-cut distinction between official and private mapping activities. By the same token, some private cartography is subsidized to such an extent by government funds that it is questionable whether it can be properly designated as unofficial. Governments, particularly national governments, are major patrons of scientific activity today, as individuals, especially rulers, were during earlier periods. In one way or another, government grants support much cartographic activity in universities and professional societies. Such institutions often have cartographic staffs that, particularly in times of national emergency, have been engaged extensively in government contract work. Nevertheless, as a matter of convenience, the chapters on modern cartography in this book have been divided between predominantly official mapping on the one hand and predominantly private on the other. Ideally, in private cartography, the researcher initiates the project, receives individual credit for it, and retains responsibility for the work, even though it may be supported by public funds. Specific credit is sometimes given for government cartographic work, but as a general rule the project is the responsibility of the department concerned. This has had an important effect on cartobibliography: previously, and ideally, an individual was credited, but today this is not always possible and a commercial company, society, or government department is listed.[3]

In addition to the maps on the pages of their periodicals, various private geographical societies provide readers with occasional folded map supplements of larger-than-page-size formats. This is a well-established tradition in Europe, and in the United States *The Geographical Review, The Annals of the Association of American Geographers,* and *National Geographic* magazine, for example, all publish map supplements. The National Geographic Society's predominantly political maps, with a great many place-names in a distinctive lettering style especially designed for the purpose, are very well known worldwide. In addition to these, the NGS cartographic output has included a large number of thematic map supplements on ethnographic, archaeological, historical, biological, oceanographic, astronomical, and other subjects.[4] In connection with "official" mapping, we have discussed the "Map of Hispanic America on the Millionth Scale" issued by the American Geographical Society. The same society published another outstanding series of maps between 1950 and 1955 showing the world distribution of cholera, dengue and yellow fever, leprosy, malaria, plague, polio, and so forth, the work of Dr. Jacques May.[5] These maps are in the great tradition of medical cartography advanced by the work of Dr. Snow and others in the nineteenth century. When the Association of American Geographers implemented a map supplement series in 1958, its objective was to provide a publication outlet for geographic data on a format larger than that of its journal, the *Annals of the A.A.G.*[6] Since that time a number of topics have been covered cartographically, including "The Kingdom of Sikkim," "[American] Indian Land Cessions," "The Maghreb—Population Density," and many others.

Frequently the sheet maps issued by private societies are individual efforts that may not relate specifically to other maps; they often represent the culmination of many months or even years of research and are frequently on themes that would not otherwise be covered cartographically. Perusal of back numbers of appropriate journals will reveal the variety, quality, and extent of such coverage. As will be appreciated, these maps perform a valuable function, since they show distributions that cannot adequately be represented on the pages of a journal and they complement, rather than duplicate, mapmaking activities of other organizations. Taken together, the cartographic output of such societies is impressive, varied, and often experimental in nature. Sheet maps illustrating demographic, ethnographic, linguistic, geological, biogeographical, medical, historical, and a wide range of other distributions have been made under the auspices of professional societies.

To illustrate this type of cartography, a small sample of a population map of California by the author appears as figure 9.1. This map, published in the map supplement series of the *Annals of the Association of American Geographers,* was produced as part of the work of the Commission on a World Population Map (later the Commission on the Cartography and Geography of World Population) of the International Geographical Union.[7] The map utilizes areal symbols and open circles, proportional to population, for cities; larger cities are keyed by letters to a list of places and their population. Dots with a unit value of fifty persons are used for rural population. The value, size, and location of dots all present special problems to the cartographer. The original of this map utilizes red (pink) tint screen for urban areas, red dots for rural population, and black circles and lettering for cities, with a buff shaded-relief background, to help explain demographic concentration and dispersal. It is an attempt to solve a central problem of demographic mapping, namely to represent on one map the great difference in population density between urban and rural areas. For this predominantly independent mapping venture, the shaded-relief base was provided by the USGS, which again illustrates the dependence of private cartography on government mapping.

In addition to the two-dimensional symbols such as those shown on figure 9.1, simulated three-dimensional symbols—spheres (developed by the well-known Swedish geographer Sten de Geer), cubes, and so forth—have also been used extensively for population and economic mapping. However, a psychometric study suggests that these simulated volumetric symbols "merely reflect their perceived area" and "are not efficient in creating the desired impression of volume."[8] A number of other cartographic symbols—graded patterns, circles, dots, and so on—

*Figure 9.1.* Small section of the map "California Population Distribution in 1960," by the author focusing on the San Francisco area (original scale: 1:1,000,000).

have been subjected to rigorous analysis, as have figure/ground relationships and color. Historically, cartographic methods to serve the physical sciences have been better developed than techniques used for the social sciences, but the latter are now receiving a good deal of attention.[9] This order of development follows the general progress of science, in which, by and large, the physical sciences reached a high level of sophistication earlier than those studies dealing systematically with life and human beings, especially psychology.

Along with this greater emphasis on the mapping of cultural features has come a greater understanding of the importance of psychometric testing of symbols. Previously it was assumed that cartographers knew how to communicate symbolically, but this is increasingly called into question as the speed of travel increases and maps on video screens are exposed for only short periods of time. In addition, maps are frequently used by the illiterate, or by those unable to read the language(s) used on a particular map. Accordingly, the nature and quality of the graphic symbols are of the utmost significance in cartography, as visual language replaces—or, at least, to a large degree supplements—verbal language. Symbols have been devised for many human activities, from agriculture to vehicle control, including geography. Actually all of the symbols used in geography are cartographic, and we have encountered many of them in the preceding pages—linear, especially roads, boundaries, rivers, and contours; point, including settlements, dwellings, and other structures; and areal, such as glaciers, lakes, and vegetation. These can be further divided into qualitative and quantitative, static and dynamic, the latter often shown by an arrow or flow line.[10] As we have seen, attempts have been made to internationalize graphic symbols, including those used in cartography, but this has not been fully accomplished and there remains a good deal of variation in symbolization, including use of color, among different mapping companies and agencies.[11]

As we travel the "information superhighway," symbolic language becomes increasingly important and internationalization of great concern. A very wide range of maps dealing with communications has appeared in recent years, but many of these are known and used only by specialists.[12] On the other hand, transportation maps are among the most familiar of cartographic works. Maps used for public transportation (bus, train, underground) are often schematic or diagrammatic and emphasize only those features, such as stopping places, of immediate concern to the traveler.[13] We have discussed the railroad map as an important cartographic genre of the nineteenth century. It continued into the twentieth century, but as other modes of transportation supplemented and then overtook the railroad for passenger traffic, new map forms

appeared. The verbal guide and the bicycle map were important at the
turn of the century, when they flourished side by side with the map designed for the motorist, who at first might venture only a few miles from
home. After a federal, numbered highway system was approved by the
U.S. Congress in 1926, local maps were superseded by maps showing the
then-new national roads and, later, interstate highways (limited-access
routes) in the United States and comparable roads (autobahn, auto-
strada, motorways) in Europe and elsewhere.

As the road map became increasingly common, its character
changed from local coverage to that of larger and larger areas, and from
more general to specifically thematic. Today, the sheet map perhaps
most familiar to the general public, especially in the United States, is the
automobile road map. In this country over two hundred million such
maps are now distributed annually, mainly by oil companies. The ancestry of the road map can be traced to the earliest cartographic efforts, but
the modern automobile map is a product of this century. State highway
commissions and automobile associations are significant producers of
road maps, and in some countries tire companies are the chief purveyors
of automobile maps. But in terms of numbers of road maps published
in the United States, the oil companies have dominated since they began
free distribution of maps in 1919. This free distribution, which was dis-
continued in the late 1970s, had a very negative effect on cartography,
as maps were used for such purposes as covering the picnic table and
were not sufficiently valued. Now that they cost on the average a couple
of dollars each, they are not considered disposable items and are
patched with mending tape to prolong their life. Also, as more markers
were erected and roads constantly improved, the symbolization on
American road maps developed into that we know today. Eventually
three private map companies, which still control much of this business
today, came to dominate automobile-map publishing in the United
States. However, recently a number of smaller local and regional companies have entered the field, producing maps and atlases (often
computer-generated) especially tailored to the needs of the communities and clientele they serve.

Although the basic symbolization for road classification and other
significant elements was initiated in the 1920s, there have been many
improvements in the design of automobile maps since that time. Color
printing and plastic scribing techniques, as well as preprinted lettering
and patterns (the latter now being largely replaced by computer methods), have been used with conspicuous success in road map production.
Figure 9.2 shows the highway symbols recommended by the American
Association of State Highway Officials and their use on a map of part of

the Chicago area. These recommendations, however, have not always been followed by map publishers, and road maps vary considerably one from the other. Not shown on this sample are such features as lists of streets with map references and coordinate systems; tables of distances between places on the map; and large-scale and small-scale inset maps. The modern American automobile map, because of constant revision, is normally the most current source of information on roads, boundaries, and places of interest in the area covered. However, road maps very often lack relief representation, and, for this and other reasons, do not take the place of topographic maps as a general source of information concerning the landscape.[14] Eventually, it is predicted, the automobile map may be largely replaced by electronic graphics; already vehicles have been produced equipped with CD-ROM road maps and Global Positioning Systems (GPS). The latter enables the operator to collect data of objects useful for mapping and navigation by utilizing free-access electronic signals emitted by satellites, especially Navstar, which orbit the earth at an altitude of approximately nineteen thousand kilometers. By locking on to at least four satellites at a time, a reliable fix (latitude, longitude, and altitude) for any point on the surface of the earth can be obtained within seconds. Accuracy to within one to five meters can be expected from individually operated, hand-held instruments. With instruments mounted on a tripod and with longer-term averaging, accuracy to within a few millimeters is possible, a situation useful for monitoring tectonic movements and in geodesy. Electronic maps for vehicles may be either linear (autoroute strip-map type) or areal, similar to the "DeLORME Street Atlas USA," which can project a map of any road in the country on a computer screen together with its ZIP (postal) code. Street addresses and individual telephone numbers for small areas can be provided electronically in some systems.

As air travel has become more and more common, airlines have been enterprising in providing interesting maps for their passengers. Outstanding in this regard is the work of Hal Shelton, who prepares "natural-color" maps of great aesthetic appeal. Shelton perfected this method of relief representation, in which color, rather than expressing altitude per se, is used to indicate the expected hue of the surface, especially vegetation, at the height of the growing season. This scheme, combined with shading, gives a remarkably graphic picture of the earth as it appears to the air traveler, and Shelton's maps have been used extensively by airlines. Such maps often have an overprint of the approximate route of the plane. These routes, of course, have no visible expression on the landscape (except for terminals and runways). Underwater cables and buried pipelines, as well as caves, which also cannot be seen on

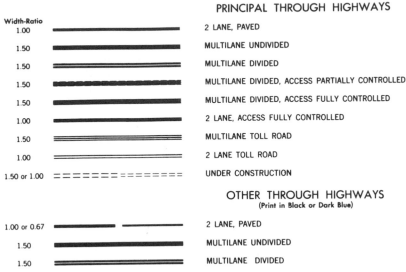

## PRINCIPAL THROUGH HIGHWAYS

| Width-Ratio | | |
|---|---|---|
| 1.00 | | 2 LANE, PAVED |
| 1.50 | | MULTILANE UNDIVIDED |
| 1.50 | | MULTILANE DIVIDED |
| 1.50 | | MULTILANE DIVIDED, ACCESS PARTIALLY CONTROLLED |
| 1.50 | | MULTILANE DIVIDED, ACCESS FULLY CONTROLLED |
| 1.00 | | 2 LANE, ACCESS FULLY CONTROLLED |
| 1.50 | | MULTILANE TOLL ROAD |
| 1.00 | | 2 LANE TOLL ROAD |
| 1.50 or 1.00 | | UNDER CONSTRUCTION |

## OTHER THROUGH HIGHWAYS
### (Print in Black or Dark Blue)

| | | |
|---|---|---|
| 1.00 or 0.67 | | 2 LANE, PAVED |
| 1.50 | | MULTILANE UNDIVIDED |
| 1.50 | | MULTILANE DIVIDED |
| 1.00 or 0.67 | | DUSTLESS |
| 1.00 or 0.67 | | OTHER ALL WEATHER |
| 1.00 or 0.67 | | UNIMPROVED |
| 1.00 or 0.67 | | UNDER CONSTRUCTION |

## OTHER HIGHWAYS
### (Print in Gray or Light Blue)

| | | |
|---|---|---|
| 0.67 | | 2 LANE, PAVED |
| 0.67 | | DUSTLESS |
| 0.67 | | OTHER ALL WEATHER |
| 0.67 | | UNIMPROVED |
| 0.67 | | UNDER CONSTRUCTION |

## ACCESS POINTS

FULL TRAFFIC INTERCHANGE

PARTIAL TRAFFIC INTERCHANGE

ACCESS DENIED

## ROUTE MARKERS

| | |
|---|---|
| 80 | INTERSTATE MARKER |
| 80 | INTERSTATE MARKER (Business Loop or Spur) |
| 42 | U.S. |
| 25 | STA |
| 68 | OTHER ROUTE MARKER |

## EXAMPLE OF RECOMMENDED HIGHWAY SYMBOLS
### IN A CONGESTED AREA
Scale 1″ = 12 ½ Miles    Width-Ratio 1.00 = ¹⁄₄₀ th Inch
(Disregard symbols for all other cultural features)

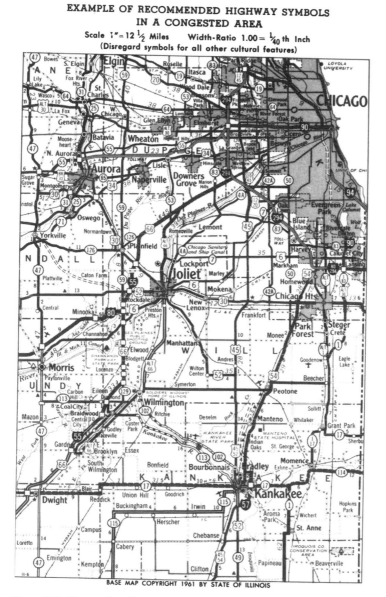

BASE MAP COPYRIGHT 1961 BY STATE OF ILLINOIS

*Figure 9.2.* Conventional symbols for highway maps recommended by the American Association of State Highway Officials.

the surface, are the subjects of special maps, sometimes of great complexity. However, such maps, like those showing communications, are often used only by engineers and other specialists who need to consult them.

Just as routes have been a feature of maps since the beginning of cartography, so has terrain representation. We have seen how, during most of cartographic history, hills and mountains have been represented in profile or three-quarter view, in contrast to the plan view of the maps on which they appear. We have discussed also the relatively late development of planimetrically correct methods of terrain rendering, such as the quantitative isobath and contour and the more qualitative hachures and shading. Physiographic diagrams and landform maps, which became popular in America in the mid–twentieth century, represent a return, geometrically, to the earlier period of cartography, even though in their richness of information the best examples are a great advance over earlier efforts at non–planimetrically correct terrain representation. Followers of William Morris Davis in this century took the surface symbols of the block diagram and placed them on a map base. The result is a useful and easily understood picture of an area. Douglas Johnson (1850–1934) and others used this technique, which reached a high level in the physiographic diagrams of Armin K. Lobeck (1886–1958), who refused to use the term *map* for these renderings because of the inconsistency of the view of the terrain in relation to the map base, and in the landform maps of Erwin J. Raisz (1893–1963) (fig. 9.3). Lobeck's diagrams were much used for didactic purposes, while the landform maps of Raisz have been employed for these as well as for book illustrations. Raisz's ability to depict landforms, in particular, could not be easily duplicated by others, and he left no substantial cadre of disciples.

The inconsistency between perspective landform drawing and the plan view of the map has been a source of concern to a number of cartographers. The basic problem, of course, is that the landform symbols are not planimetrically correct, which becomes progressively more of a problem as the scale of the map increases. In an effort to illustrate the brightness of the configuration of a surface, Kitirô Tanaka, a Japanese cartographer/engineer, developed and utilized traces of parallel inclined planes, or, as he called them, "inclined contours."[15] By a rather simple technique, Tanaka transformed horizontal contours to the traces of the intersection of parallel, inclined planes with a surface. The traces used by Tanaka resulted from the inclination of planes at 45 degrees; figure 9.4 shows the relationship between the horizontal contour, the profile, and the trace of an inclined plane. As recognized by Tanaka and others who commented on the technique, maps drawn with a series of

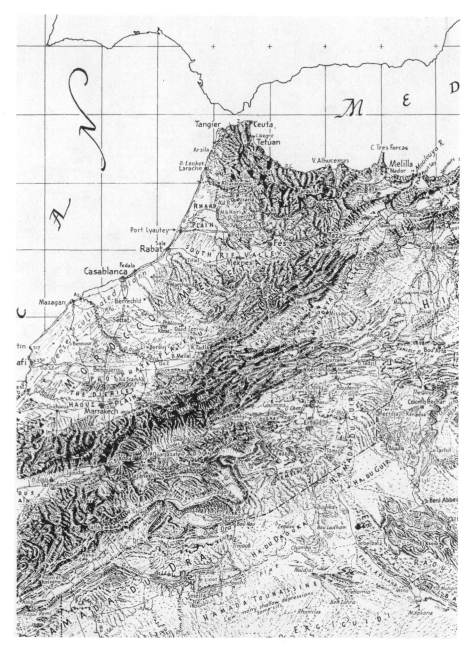

*Figure 9.3.* Small part of landform map of North Africa by Erwin J. Raisz.

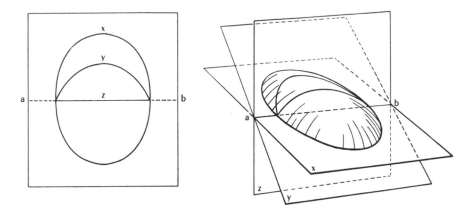

*Figure 9.4.* Two- and simulated three-dimensional forms of the relationship between a contour (x); a trace of an inclined plane (y); and a profile (z).

traces of parallel planes inclined at 45 degrees have a number of objectionable features. Among these are (1) landforms that appear too flat; (2) an overall darkness to the rendering, which makes overprinting difficult; and (3) a pseudoscopic (inverted-image) effect. In the mid-1950s, Arthur H. Robinson and the author experimented with traces of planes having angles other than 45 degrees with the horizontal.[16] They used the traces to define forms and discovered that whatever the angle used, correct planimetry was retained. By using angles of less than 45 degrees with the datum, the first two objections above were overcome, while the third was resolved by drawing fewer traces on the side facing a hypothetical light source or, conversely, heavier lines on the side away from this light source. The result is a planimetrically correct method of terrain rendering that permits the preparation of effective landform-type maps of a larger scale than were, in practice, possible previously (fig. 9.5, drawn by the author). Such maps can be overprinted with information taken directly from vertical aerial photos and from other maps. This method is much more objective than the physiographic method as well as easier for students to learn, and it has been computerized.

The appearance of the third dimension in cartography has been achieved in various ways in addition to those discussed above. We have illustrated shaded relief on a topographic map; this technique has been used in a remarkably effective way at geographical scale in maps made by Richard E. Harrison.[17] Harrison (1901–94) was a trained artist and architect who became interested in relief representation and drew per-

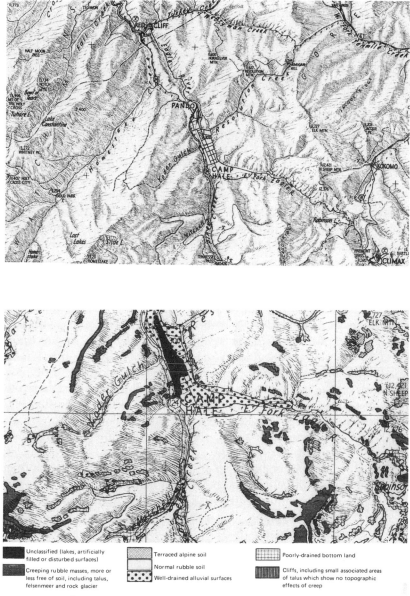

Because terrain types intergrade and mingle, boundaries on this map separate areas of dominance

Legend:

- Unclassified (lakes, artificially filled or disturbed surfaces)
- Creeping rubble masses, more or less free of soil, including talus, felsenmeer and rock glacier
- Terraced alpine soil
- Normal rubble soil
- Well-drained alluvial surfaces
- Poorly-drained bottom land
- Cliffs, including small associated areas of talus which show no topographic effects of creep

*Figure 9.5.* Above, central portion of a relief map of part of the
Colorado Rockies drawn by the author using traces of inclined planes as a basis
(original scale: 1:50,000). Below, large-scale section of the map above,
with overlay of selected surface information.

spective-type terrain renderings at very small scales. Many of these are on orthographic, perspective projections, showing landmasses on the globe from unusual orientations, but much of his later work was in plan view. An example of the latter is the submarine relief of the Atlantic Ocean illustrating, among many other features, the Mid-Atlantic Ridge (fig. 9.6); the original is in color, and much is lost in the monochrome reproduction. There is a great tradition of terrain mapping in Alpine countries, as mentioned earlier. Swiss cartographers, notably Eduard Imhof in the twentieth century, have produced splendid terrain representation employing shaded relief and aerial perspective, in which warmer colors are used for higher elevations and cooler ones for lower areas.[18] Photographic and contour-line anaglyphs (composite stereoscopic images) have been employed with great success in the French *Relief Form Atlas*.[19] Though hardly classified as maps, a large variety of raised (tactile) globes, parts of globes, and terrain models is now available. At one time these were usually made of plaster, often painted in realistic colors, but now other materials, particularly plastic, are used. These new materials have made the raised model lighter and easier to manufacture but have not solved the great storage problem associated with such visual aids: many maps can be stored in the space required for a single model. Another major problem with raised models, and indeed with profiles, is the degree to which these devices should be exaggerated in the vertical dimension to give a realistic appearance.[20] Models and profiles tend to look too flat unless vertically exaggerated because one is apt to think of elevations on the land as being relatively higher than they are in reality. A tactile surface has a very special use in maps for the blind that have been designed and produced in recent years. This is in line with the general emphasis on making places such as parks and buildings more accessible to the handicapped.[21]

The third dimension, whether simulated or actual, is not confined to representation of landforms, and other map data can be shown in this manner. We have illustrated statistical surfaces in two- and simulated three-dimensional form (fig. 7.13), the traces of inclined planes have also been used for population, rainfall, and other distributions.[22] Oblique-view maps of urban areas in the tradition of those of the Renaissance (fig. 5.12) have recently been made by, among others, Hermann Bollmann, a German cartographer, and his associates. After creating bird's-eye-view maps of a number of European cities, Bollmann turned his attention to New York. His representation of that city must be among the most remarkable maps of modern times, showing, as it does, buildings and other features with great accuracy and detail (fig. 9.7). Typically, modern urban geography deals with central-place theory, the functional areas of the city, the reach of wholesale and retail centers, business land use, ethnic areas, health, poverty, wealth, and crime, all of which have been mapped for major metropolitan centers. Through such maps, a great deal of valuable information is available to planners and others, but these maps, as well as conventional urban plans, do not convey the visual impact of the city in the same way as do isometric or perspective maps. Artists such as Bollmann, Shelton, and Harrison produced maps in the mid–twentieth century that are artistically the equal of the best work of cartographers of the past. Furthermore, the techniques by which their maps are reproduced are vastly superior to earlier methods, which were incapable of bringing out all of the subtlety of original artwork.

Maps showing countries in human or animal form are part of a venerable tradition, of which Leo Belgicus, the Lion of Flanders, dating from the early seventeenth century, is a prime example. Other "cartographical curiosities" include maps as playing cards, board games, jigsaw puzzles, and the subject of cartoons.[23] Maps reproduced in magazines and newspapers reach a large number of people who might otherwise be little concerned with cartography. The famous Gerrymander cartoon-map appeared in the Boston *Gazette* in 1812, and maps have been a feature of American journalism ever since. Subjects covered include war and political events, travel and recreation, and, of course, the weather (treated earlier). Journalistic cartography is familiar to almost everyone, and it ranges from sophisticated and highly original maps in color for magazines printed on high-grade paper to simple line drawings in black and white that reproduce satisfactorily on soft newsprint.[24] In the former category are the maps of Harrison for *Fortune* magazine. Like those of Shelton, these maps are interesting from a technical point of view as examples of cartography in which the color separation is normally accomplished by the photographer from the finished artwork (four-color process) rather than by the cartographer, as is usual with topographic and most other colored maps (flat color). Maps showing parts of the world from unfamiliar viewpoints by Harrison, and those of the Pacific area by the Mexican artist Miguel Covarrubias, originally published in *Fortune,* have become classic examples of their respective cartographic types. At the other end of the scale are local newspaper maps, which are usually simple line drawings reproduced in black and white, often drawn by a staff artist with no training in cartography.

It is a short step from journalistic to persuasive cartography, in which maps are designed to "persuade" or change the opinion of the viewer. This term was coined by Judith Tyner to include, among others, maps used by theologians to illustrate current beliefs, by advertisers to sell products, and by propagandists to mislead the enemy.[25] Tyner recog-

*Figure 9.6.* Submarine relief of the Atlantic Ocean by Richard E. Harrison showing the Mid-Atlantic Ridge, continental shelves and scarps, abyssal depths, and so forth.

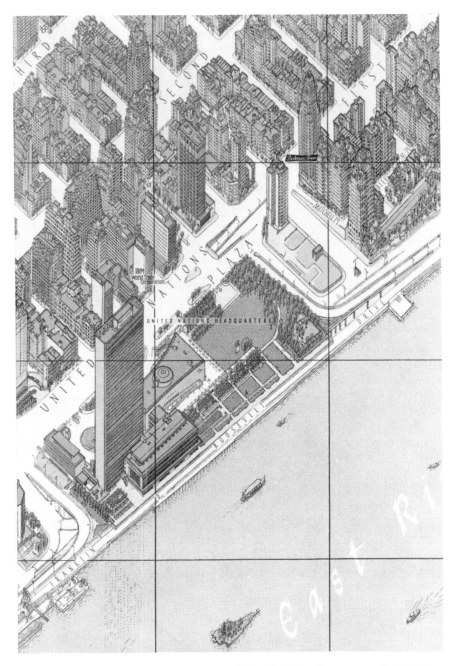

*Figure 9.7.* Section of an isometric map of New York City focusing on the United Nations Plaza by Hermann Bollmann (slightly reduced).

nizes deliberate, unintentional, and reader-caused persuasive cartography and cites Nazi propaganda maps of World War II, improper use of the Mercator projection for world distributions, and Sir Halford J. Mackinder's map of the "heartland" as examples. Mackinder, by marginalizing the Americas on his political world map on an ovoid projection, may have misled himself. In persuasive cartography, the map itself becomes a psychological tool. A whole arsenal of devices may be used for such mapping: color; projection; maps in series; size relationships; symbols, especially arrows; and boundaries. These are employed to influence opinion rather than to dispassionately inform the map reader. The authority of the map and globe, which is emblematic of and synonymous with education, is invoked, even though some persuasive maps contain intolerable errors. For example, the author once saw an advertising map of the Americas in which the equator passed through Panama rather than Brazil and Ecuador, probably because it seemed a "natural" place for 0 degrees latitude. Then there is the "four-polar" projection used by a Los Angeles supermarket chain to help encourage recycling (fig. 9.8)! We should be properly critical of all media output—including printed maps.

*Figure 9.8.* The "four-polar" projection used in advertising!

Of course, all maps are to a greater or lesser degree "subjective," and those that consciously distort from an areally "correct" shape have become increasingly common in recent years; the possibilities are limitless. They range all the way from interpretations of individual perceptions of phenomena that may be only in the mind and, if expressed in some manner, may or may not be intelligible to anyone but the originator, to such well-known examples as "The New Yorker's Idea of the United States of America," the humor of which can be appreciated by millions of people.[26] In this map, the sizes of places in the United States are in relation to their assumed importance to the inhabitants of New York City, so that Manhattan is considerably larger than the state of Washington (which, incidentally, is located south of Oregon). The size of an area, however, is not simply a function of distance from New York, for Florida appears larger than all the states intervening between it and New York. The so-called Area Proportional To (APT), or area cartogram principle, has also been used for serious purposes, as in Pierre George's well-known cartogram showing the size of countries of the world according to population.[27] The same concept has been used for the map "The Political Colour of Britain by Numbers of Voters" published in *The Times* of London (19 October 1964), in which "(1) areas of the map are proportional to the population and not to the amount of land, (2) contiguity of areas is everywhere preserved, and (3) the relative disposition of places on the map is as close as possible to the relative geographic location." A quarter of the bulk of the map is in the populous southeast, around London, while the north of Scotland is much smaller than its considerable geographical size would suggest. To illustrate this principle we reproduce three maps of Hungary adapted from a study of economic maps by Lászlo Lackó (fig. 9.9).[28] The maps are all drawn within the framework provided by the geographical limits of the country, but the area of the internal political units differs proportionally according to the phenomenon being mapped: from top to bottom, geographical area, population, and industrial employees. Geometries other than the Euclidean and coordinate systems other than the Cartesian are increasingly used in cartography.

Area is not the only geographical quality or element that can be altered to suit particular purposes. For example, on a much-reproduced map by Torsten Hägerstrand, distances from the parish of Asby are scaled logarithmically to provide a more useful cartographic base for plotting migration flow, telephone calls, and other distributions.[29] Among European cartographers, the Scandinavians—particularly the Swedes—enjoy a reputation for originality in statistical mapping equal to that of the Swiss in terrain representation. The economic maps of W.

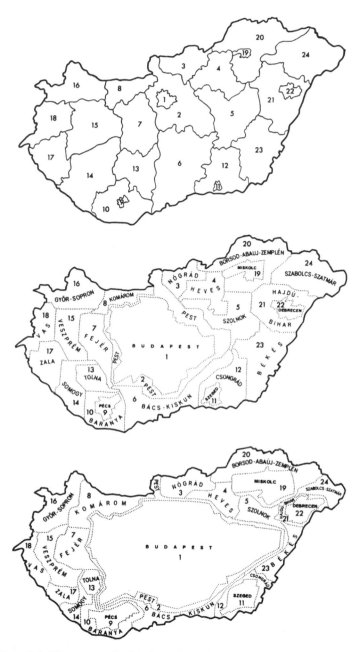

*Figure 9.9.* Three maps by László Lackó showing the size of the counties of Hungary (*from top to bottom*): according to geographical area; in proportion to population; and in proportion to the number of industrial employees.

William-Olsson, for example, are extremely informative and contain a number of cartographic innovations. Problems associated with economic mapping over large areas of Europe are much greater than in the United States because of the lack of uniformity in the census reports of the several countries involved. Nevertheless, attempts have been made to reduce the information to a common base, as in William-Olsson's "Economic Map of Europe" of the scale of 1:3,250,000 and in the *Atlas of Western Europe* by Jean Dollfus.[30] As the European Community (EC) broadens its membership and becomes more than an economic union, we will see greater opportunities for multinational cooperation in many areas, including cartography. Spearheaded by a private American map company, a thoroughly international atlas of the world has been published.[31] This was developed in cooperation with similar cartographic organizations in Hungary, Sweden, Britain, Germany, and Japan. A very valuable feature of this atlas is the consistent use of a few scales for certain classes of maps, including plans of all major metropolitan areas of the world on the scale of 1:300,000. A problem that almost defies solution in such a work is the rendering of place-names on the maps, which in the case of *The International Atlas* are gazetteered according to the local name and in English, German, Spanish, French, and Portuguese. On the maps in the *Atlas,* names of important places may appear in more than one language (Livorno, Leghorn).

Publicly displayed maps are a familiar sight in various cities and towns, at bus and railway stations, and so forth.[32] Sometimes they are made of permanent materials such as glazed tiles, but more often they are printed maps enclosed in a case. Often the map is pictorial in character, especially if it is of a smaller tourist center. In recent years guide maps of a rather elaborate type have been placed in strategic locations, especially in European cities, housed in substantial, glass-fronted installations. Frequently the map is mounted on cloth and attached to rollers so that by turning a knob the reader can raise or lower it to a desired height. A classified register of places of interest to the tourist and an alphabetical list of streets with map references are typical features of such display maps. The permanently installed, vertically mounted map is a device of utility to city visitors, but special types, some of which are electronically controlled to display needed information to businessmen, clients, and others, are also used in offices. A remarkable product of this general order is the Geochron, which, as its name suggests, gives the time at any place on the earth. This is accomplished by an illuminated Mercator projection that continuously records the pattern of light and darkness (the circle of illumination on the globe) as well as the azimuth (the point where the sun is directly overhead) and time. The shape of the circle of illumination changes on the Mercator projection from bell shape (winter solstice) to inverted bell (summer solstice) to vertical (equinoxes) as the year progresses. A step in this direction is a globe that automatically rotates once during twenty-four hours.

We have seen how through the centuries the problem of transferring the spherical grid (graticule) from a globe to a plane surface has been solved. We have also noted that, although all global properties cannot be preserved in any one projection, these devices may possess positive qualities that make them more than poor substitutes for the globe. Recent advances have been made to this study by the development of new projections, adaptations or new cases of existing projections, and new uses for projections devised in earlier times. Relatively few individuals have invented original, useful projections, but all cartographers should know enough about projections to make good choices among those that exist. Indeed, the nonspecialist who looks at maps intelligently ought to know something of the strengths and limitations of the projections used (appendix A).

Various attempts have been made to classify projections, one profitable way being in terms of geometrical shapes that may actually or theoretically be used for their construction.[33] Many shapes may be employed for this purpose; three commonly used are cylinders, cones, and flat surfaces, giving rise, respectively, to cylindrical, conic, and azimuthal classes of projections (fig. 9.10).[34] Thus one may have a cylinder touching a generating globe along one line or intersecting to touch along two lines, therefore spreading the distortion or deformation (such a line of contact is known as a standard line and is often, but not necessarily, a parallel); a cone touching along one line or intersecting to touch along two lines; a flat surface in contact at one point, or intersecting, or at a distance from the generating globe. One may further conceive of a transparent globe (or hemisphere) on which are drawn opaque lines of latitude and longitude and a light source with which to "project" the grid onto the shapes described above to produce projections with different qualities. For example, a light source at the center of the globe projecting onto a flat surface in contact with a globe at one point produces a gnomonic projection; if the light source is on the side of the globe exactly opposite the flat surface, a stereographic projection results; if the light source is at an "infinite" distance, the orthographic projection is produced. All of these projections, devised centuries ago, have found new and important uses in the modern world of rapid transportation, communication, and global thinking.

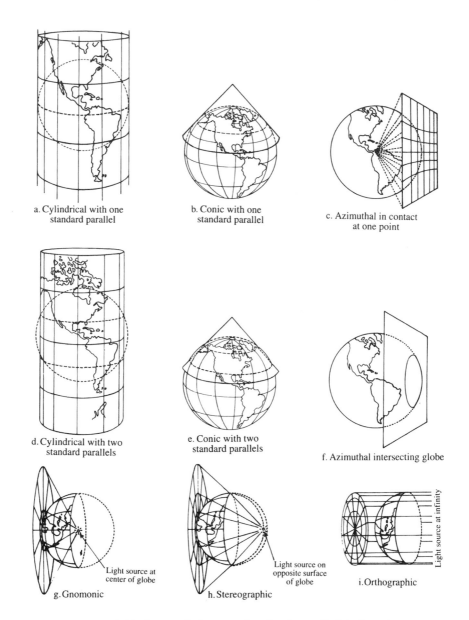

a. Cylindrical with one
standard parallel

b. Conic with one
standard parallel

c. Azimuthal in contact
at one point

d. Cylindrical with two
standard parallels

e. Conic with two
standard parallels

f. Azimuthal intersecting globe

Light source at
center of globe

g. Gnomonic

Light source on
opposite surface
of globe

h. Stereographic

Light source at infinity

i. Orthographic

*Figure 9.10.* Geometrical properties of various cylindrical, conic,
and azimuthal projections.

Of course not all earlier projections fit into this simple classification. Neither do a number devised in the twentieth century, including the circular projections published by Alphons J. Van der Grinten (1852–1921) in 1904 (I and IV) and 1912 (II and III). The Roman numerals were applied later to differentiate variants. Van der Grinten I (fig. 9.11a) has meridians and parallels that are circular arcs (except the equator and central meridian, which are straight lines). On Van der Grinten II, the parallels and meridians cross at right angles; on III, the parallels are straight lines; and IV is apple-shaped. Only Van der Grinten I has been in regular use, employed by the National Geographic Society for its emblem and for many of its world maps from 1922 to 1988. At the latter date, the NGS replaced the Van der Grinten with the Robinson projection (fig. 9.11g). Arthur Robinson had devised this projection, which is a pseudocylindrical, compromise arrangement of the earth grid, in 1963.

Like Van der Grinten, another early-twentieth-century cartographer who sired a family of projections was the German scholar Max Eckert (1868–1938; after 1935, known as Max Eckert-Griefendorff).[35] In 1906 Eckert published six pseudocylindrical projections, which are named after him; all of these projections are alike in that parallels are represented by straight, parallel lines and the poles are half the length of the equator. They differ, however, in that Eckert I and II have straight meridians (broken at the equator), III and IV have elliptical meridians, and V and VI sinusoidal meridians. The spacing of the parallels of II, IV, and VI is such that they are equal-area (equivalent) projections, while the other three are not. Eckert IV has proven to be the most popular of the six (fig. 9.11b).

We have seen the characteristic of meridianal interruption, a technique to improve the shape of the representation, in the gores of the globe and alluded to it in connection with the (interrupted) double cordiform projection of Fine and Mercator. In 1916 J. Paul Goode, a geographer at the University of Chicago, applied this principle to the sinusoidal and Mollweide projections.[36] As shown in figure 9.11c, Goode used a series of lobes, each with a standard meridian but hinged along a common equator, on the Mollweide (homolographic) projection. He applied the same method in 1923 to his so-called homolosine projection, a combination of the lower-latitude section of the sinusoidal (0–40 degrees north and south) with the higher-latitude parts of the Mollweide (40–90 degrees north and south). In the homolosine, Goode grafted the "best" parts of two well-known equal-area projections and further reduced distortion by interruption (fig. 9.11d). A further modification of the grid undertaken by Goode involves condensation, in which areas of little interest are deleted (the shaded area of fig. 9.11d) and the remaining areas are moved together to save space. In all maps, whether simple or complex, an indication of the grid should be given either by continuous lines or by crosses and ticks so that the geometry of the projection used may be understood.[37] An equal-area projection that is the arithmetical mean between the Mollweide and the sinusoidal was developed by S. Whittemore Boggs (1889–1954) in 1929. This is known as the Boggs eumorphic (equal-area) projection.

During World War II other projections were created, such as that devised by Osborn M. Miller (1897–1979) a British worker at the American Geographical Society of New York. The Miller cylindrical projection forms a rectangle, as does the Mercator, but unlike that projection it allows the poles to be shown. It avoids the extreme deformation of the Mercator in high latitudes and the great angular exaggeration of the cylindrical equal-area projection and, therefore, is a useful compromise for displaying the whole earth (fig. 9.11e).[38] Cylindrical projections have been used and misused during the century, including the Gall Orthographic, which was "reinvented" in 1973 by Arno Peters, who made extravagant claims for it, including its ability to show Third World countries more fairly than so-called "Eurocentric" projections such as the Mercator. What amounted to a proscription against the improper use of cylindrical projections was issued by the American Cartographical Association and others in 1989. Oblique and transverse cases of well-known projections discussed briefly in chapter 7 have been used increasingly in recent times. In the transverse (polar) case, the grid is shifted 90 degrees so that, for example, the transverse Mercator is tangent upon a meridian rather than the equator, as in its standard or conventional form. If the shift is less than 90 degrees from the normal position, the term *oblique* is used. In its transverse and oblique cases, the Mercator remains conformal (grid lines cross at right angles and shapes around points are correct), but its ability to show all straight lines as rhumb lines (lines of constant compass direction) is not retained. Although any projection can be shifted in the manner described above, in practice only a few are normally so treated, including among equal-area projections the Mollweide and among conformal ones the Mercator. Oblique projections are often used for world maps in modern atlases, while transverse projections, especially the Mercator, have gained increasing popularity for the topographic map series of several of the great surveys of the world (including Britain's Ordnance Survey).

As has been stressed previously, the globe is the only true representation of the earth, but this can be approximated by gores spread out on a

a. Van der Grinten I 1904

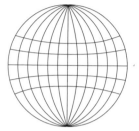

b. Eckert IV (pseudocylindrical) 1906

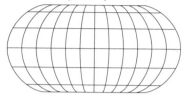

c. Mollweide (homolographic) c. 1800
   Goode (interrupted) 1916

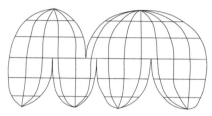

d. Goode homolosine (interrupted) 1923
   shaded areas could be eliminated allowing
   compression for certain purposes

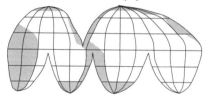

e. Miller cylindrical 1942

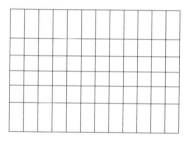

f. Fuller Dymaxion 1943
   (correct scale along fold lines--dashed;
   makes a globe of 20 equilateral triangle)

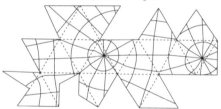

g. Robinson (pseudocylindrical) 1963

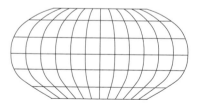

h. Perspective (orthographic) 1988 as seen by GOES
   (Geostationary Operational Environmental Satellite)

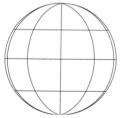

*Figure 9.11.* Twentieth-century projections; a 30-degree grid is used in all cases.

flat surface. An ingenious modern example of this is the earth (showing continental outlines) and tectonic globes of the cartographer Tau Rho Alpha, with others.[39] Alpha drew gores that can be colored, cut out, and pasted onto (regulation-size) tennis balls. The faces of near-globular geometrical solids have been used to simulate the globe (including the cuboctahedron and the icosahedron) in the Dymaxion Airocean World maps of R. Buckminster Fuller, inventor of the geodesic dome. These polyhedrons can either be used on a plane surface to show interesting relationships or folded into a solid, approximating the globe (fig. 9.11f). With the advent of the air and space ages, there has been an increasing interest in orthographic and perspective projections (fig. 9.11h). For example, a hemispheric perspective projection is used to display weather patterns as imaged by the GOES satellite and shown on television, often in animated form. A new projection inspired by the U.S. space program is the space oblique Mercator (SOM), proposed by Alden P. Colvocoresses and developed by John P. Snyder.[40] On this projection, which takes into account earth rotation, the changing groundtracks of Landsat orbits can be plotted effectively.

An idea related to map projections that deserves more attention than it has received is the grid formed by meridians and parallels for the comparison and measurement of area, proposed by Harry P. Bailey.[41] Bailey, a climatologist, used meridians and parallels to delimit areas of equal size on the surface of the earth. To accomplish this, meridians remain at a constant and equal angular interval while parallels are spaced in such a manner that quadrilaterals of uniform size on the globe are enclosed. This has the effect of producing an area-reference grid of considerable utility. Bailey applied this principle to a variety of projections—equal-area, conformal, and compromise types.

In the first edition of this book (1972), computer cartography, then in its infancy, was described and illustrated (fig. 9.12). It was mentioned that at the time the computer was more advanced in its information than in its design capability, but it was expected that this latter aspect would soon be improved. That this has indeed come about in a couple of decades can be appreciated by comparing figure 9.12 and figure 9.13. These maps, which are on different topics and at different scales, cover (in part) the same area, yet most people find the later example more appealing aesthetically as well as more informational than the earlier one. Rather than taking the art out of *cart*ography, the computer can produce effects and perform mapping tasks that were previously undertaken only with intense labor or were not even possible. By such means cartography has been transmogrified into geographical information systems (GIS).[42]

While recognizing that the computer has revolutionized the practice of cartography, in keeping with the theme of this book we will be more concerned with its effect on the product than with how computer maps are made. Thus our discussion on this aspect will be brief, and the interested reader should consult specialized texts on the topic.[43] Of course the computer did not come into being in the second half of the twentieth century ready-made, without precedents. The abacus, a manually operated storage and calculating device, has been used in the Orient for 5,000 years. In Europe, especially in the seventeenth century, a number of mathematicians were interested in computing by mechanical means. John Napier, the inventor of logarithms, described a method of performing multiplication and division by use of rods or "bones," a proto–slide rule, in 1617; Blaise Pascal built a digital calculating machine in 1642; and Gottfried Wilhelm Leibnitz (1646–1716) completed a more advanced version based on Pascal's principles in 1694 using the binary, rather than the decimal, system. A model of Leibnitz's machine was exhibited at the Royal Society in 1794, and an improved model capable of division, subtraction, multiplication, and addition was available by 1820.

This anticipated by thirteen years the steam-powered Analytical Engine of Charles Babbage (1792–1871), which used punched cards (adapted from those used for the Jacquard loom). One set of cards provided the numbers and the other the sequence of operations. Ada, Countess of Lovelace (1815–52), the daughter of the poet Lord Byron, wrote "Notes," what amounts to the first computer program, for Babbage in 1843.[44] Computing became practical after Herman Hollerith (1860–1929) combined punched cards with the then-recent electromagnetic inventions to make it possible to count and classify the 1890 U.S. Census returns, a task accomplished in a much shorter time and in much greater detail than by previous methods. But the machines remained large and clumsy through the 1950s and were mostly concerned with tabulating numbers. Graphic displays were made possible after that time by the electronic typewriter and the line printer.[45] For example, the then-current Synagraphic Mapping System (SYMAP) was capable of composing spatially distributed data in map, graph, and other forms of visual display. In making such a map, the steps normally taken were the following: (1) Coordinates of the controlling points were established on a manual digitizing board; (2) the information was entered on special coding forms; (3) information from the coding forms was transferred to punch cards; (4) the deck of cards was then fed to, and processed by, a mainframe computer (an operation that usually took only seconds); and (5) the map was printed out line by line at rates such as forty lines per second, with an average printout taking about one minute.

In comparison with the above, in creating the simple line drawings for the illustrations of projections in this book (such as fig. 9.11), the following steps were taken: (1) Source materials were scanned; (2) the resulting image was projected on a personal computer screen; (3) enhancement and other manipulations were accomplished; (4) using a graphics program, the scanned images were digitized; (5) text was added from a large number of available type styles; and (6) the graphic was printed out in approximately fifteen seconds. The advantages of computer cartography over conventional mapping include the ability to create uniform linework, to change scale easily, to facilitate editing and correction, to implement changes in design as the map is being made, and to provide almost instantaneous hard copy. Tonal screens or color can be used and percentages adjusted to give subtle (or gross) contrasts. At a more advanced level, the use of computer-driven color display screens, pen plotters, and laser printers (both monochrome and color) permits experiments in computer "art" to produce special effects, some of which have cartographic applications not attainable by other means.

Although there were many developments in the twenty years between SYMAP and later programs, a turning point was the 1982 introduction by Environmental Systems Research Institute (ESRI) of ARC/INFO, a geographical information software package that combines traditional automated cartography with advanced spatial data-base handling capability. This is accomplished by combining a series of layers, each with a different theme: relief, roads, point symbols, and labels (lettering). Specifically, ARC/INFO uses both vector (line) and raster (tabular) storage; transformations can be undertaken and questions asked concerning numbers, distance, addresses, and so forth. Mapping is accomplished through addition and subtraction, leading to the production of high-quality maps. The array of cartographic products using such a system is limitless and of potential utility to all who are concerned with geographical distributions. In addition, time saved in production can be enormous.

Use of tapes and disks permits ready storage and retrieval and greatly facilitates the manipulation of large data banks. The low cost and high speed of creating maps, made without benefit of hands, are valuable facets of computer cartography, as are its flexibility and its objectivity. The ability to make many different maps with ease from one data set and to quickly compare them with one another and with those made from other data sets is obviously a great advantage in understanding the complexity of areal distributions. Furthermore, the reliability of the original data can be evaluated by means of computer mapping. The computer is capable of producing maps according to contour (isoline), choropleth (conformal), and point (proximal) methods. Simulated three-dimensional representations can be produced by the computer with co-ordinates supplied in plan or elevation form. Similarly, map projections can be easily transformed from one case to another on a computer. Maps in series—for example, showing population change in an area through time—can be composed in three-dimensional models. In fact, the whole problem of understanding surfaces, whether real or abstract, concrete or imaginary, is greatly facilitated by the computer. Many of the earlier problems of symbolization and design in computer graphics have now been overcome: the straight-line character of curving lines; illogical value and textural progressions; the small size of the printed map; and the lack of color. With the improvements in these and other aspects, it can be said that the computer has transformed cartography as much as (or more than) other technological developments, including printing and aerial photography, changed the course of mapmaking at earlier pe-riods.

Figure 9.13 illustrates the seismic situation one week after the devas-tating Northridge, California, earthquake of 17 January 1994.[46] The epi-center of the quake, which measured 6.8 on the Richter scale, is indi-cated by a triangle (red on the original) with aftershocks in density classes shown by circles of variable size and color (red to yellow). Let-tering is in yellow and roads in black on a background of terrain varying from white (illuminated side of the mountains) to black (shadows) and gray (plains). Freeways and principal roads are designated by shield sym-bols with appropriate numbers. With this and other maps, almost all problems associated with the quake were addressed by GIS, including damage assessment, police response, relief efforts, and level of federal and state aid. Assisted by GIS, what would otherwise have taken months was accomplished in days.

We have dealt with the map as a snapshot representing phenomena at a given time. We have also suggested that a series of maps of the same phenomena at different times can be used to suggest temporal change. Furthermore, in the previous chapter it was indicated that animated dis-plays of weather from satellites are now commonplace on television news programs and used for forecasting and for research purposes. A step in the direction of animation is the time-lapse movie, in which the hiatus is short enough so that a mental idea of the preceding image is retained but long enough so that no real impression of movement results. A true motion picture differs from a time-lapse presentation in that, in the for-mer, a continuous appearance of motion is achieved. This is accom-plished by presenting the different images so rapidly that the eye is un-able to detect the short interruptions between the successive frames of

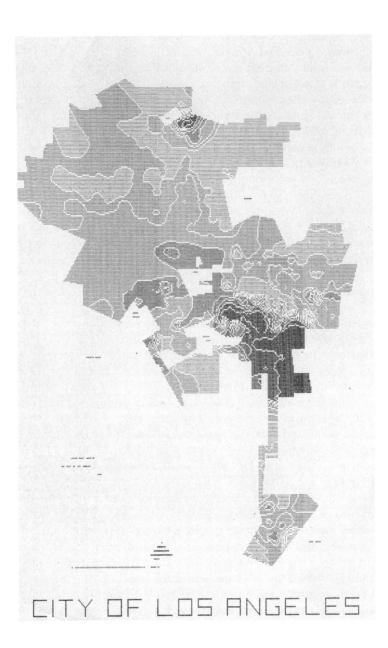

CITY OF LOS ANGELES

*Figure 9.12.* Computer map showing voting patterns in the city of Los Angeles in the 1969 mayoral election (reduced).

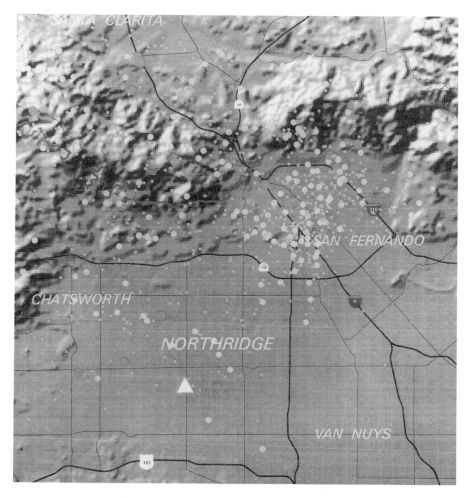

*Figure 9.13.* Detail of computer-generated map "Northridge Earthquake Epicenters as of January 24, 1994," showing the earthquake's magnitude (five classes), and fault line as well as streets and terrain.

the film. The illusion of movement thus produced, known as the *phi phenomenon,* is familiar through its use in advertising signs as well as in motion pictures. In sound films, sequential frames projected on the screen at the rate of twenty-four per second are synthesized by the eye, and the subject appears to move.

This principle has been used in very significant ways in cartography, adding the fourth dimension, time, to maps.[47] It is probable that more people now see maps projected on television and movie screens (even if these are not always animated) than any other type of map. Different mass media utilize maps for a variety of purposes, including weather reports, news programs, and entertainment, and some of them are abysmally poor examples of cartography. However, the following remarks will focus on animated cartography in educational motion pictures. Without going into detail concerning such methods, it can be said that all of the major techniques of making motion pictures have a cartographic counterpart. Thus a "live" movie can be made of a person actually drawing a simple map; this has the pedagogical advantages of a blackboard drawing, in which the image develops before the viewer. A variation of this technique is to prepare a movie of information being added to an existing map, as, for example, the route of an explorer. Another method is to create regular animated sequences in the manner of the cartoon, using a series of overlays or "cels." The same background map can be used throughout a sequence, but the transparent cels each bear a slightly different image. The cels are photographed one by one over the background to form the individual frames of a movie. When these are assembled in order and projected on the screen at the correct speed, an animated map results. Computers have now almost entirely supplanted the use of cels in the preparation of animated films, which greatly reduces much of the laborious process of making such presentations.

Thousands of movies have been made in which some map sequences appear. Symbols used in animated cartography include dots for locations; arrows for directions; lines for transportation, communications, and boundaries; pictorial symbols for agricultural or mineral products; and areal symbols for population, forests, and so on. Lettering, like the other symbols, can be added at the appropriate moment and removed when no longer needed, a very great advantage in animated cartography because it eliminates clutter on the map. A number of elements considered essential in static cartography are often omitted on animated maps. Because of the variable size of the projected image, a verbal scale or representative fraction should not be used, but a graphical scale that enlarges and reduces with the image is, of course, perfectly satisfactory. One drawback is that the animated map does not permit prolonged study unless the projector is halted, a possibility facilitated, of course, by interactive systems. For illustrating the dynamics of certain areal relationships, animation is unmatched. Animated cartography can most effectively unite the temporal element of history with the areal view of geography.[48]

To illustrate animated cartography, thirty-five small maps showing the daily progress of Christopher Columbus on his first trans-Atlantic voyage are reproduced as figure 9.14. This presentation, "Columbus' First Voyage," was made as part of a curriculum package for students in a National Endowment for the Humanities (NEH) Summer Institute, "Columbus: The Face of the Earth in the Age of Discovery" held at UCLA, 15 July to 23 August 1991.[49] The animation shows and the accompanying narration recounts the events of this momentous voyage. In the language of the programmer, the artwork was scanned with an Apple graphic scanner, reduced in size with painting software, imported into HyperCard, and edited; then the *Diario* (Columbus's log) entries were imported, and a concordance was created. The film can be projected so that it has a correct time scale, and it can be stopped at any time so that the *Diario* transcription can be consulted for any particular day. This simple idea can, of course, be elaborated ad infinitum. Interactive videos now exist that allow maps to be linked not only to a text but also to a wide range of other graphics.[50] Cartographers have not been as forward-looking as planners and architects, who have created the constantly variable three-dimensional model. This tool, analogous to the flight simulator for pilot training, allows the user to "be and walk" inside a house or city. Makers of maps could apply this technology to create perspective in physical landscapes, of value to many users.

In this survey of the development of cartography, we have seen how the map has conveyed ideas cultural and scientific, legal and political, anthropological and medical, and many others, and how it has been used in times of both peace and war. Philosophers and physicists, princes and presidents, poets and painters, physicians and priests, professors and pirates, planners and psychologists, perceptionists and programmers have in some fashion been concerned with the map. Beginning before the written record, cartography runs as a thread through history. Maps have been made by so-called primitive as well as by sophisticated peoples. By no means have professional cartographers been the only substantial contributors to this art and science; rather, because of its eclectic and universal nature, it has drawn its practitioners from many fields. Its special relationship with geography has given cartography focus, but it must not be thought of as the handmaiden of this or any particular discipline. It exists separately, even while being of interest to both the layperson and the specialist. The explosion of knowledge in very recent times is reflected in cartographic diversity and output.

Many of our most important achievements—from philosophical considerations on the nature of the earth to setting foot on the lunar surface—have had cartographic expression and, in turn, have been advanced by cartography. Over the centuries cartography has developed

**Columbus' First
Atlantic Voyage Outward
8 Sept. – 12 Oct.
1492**

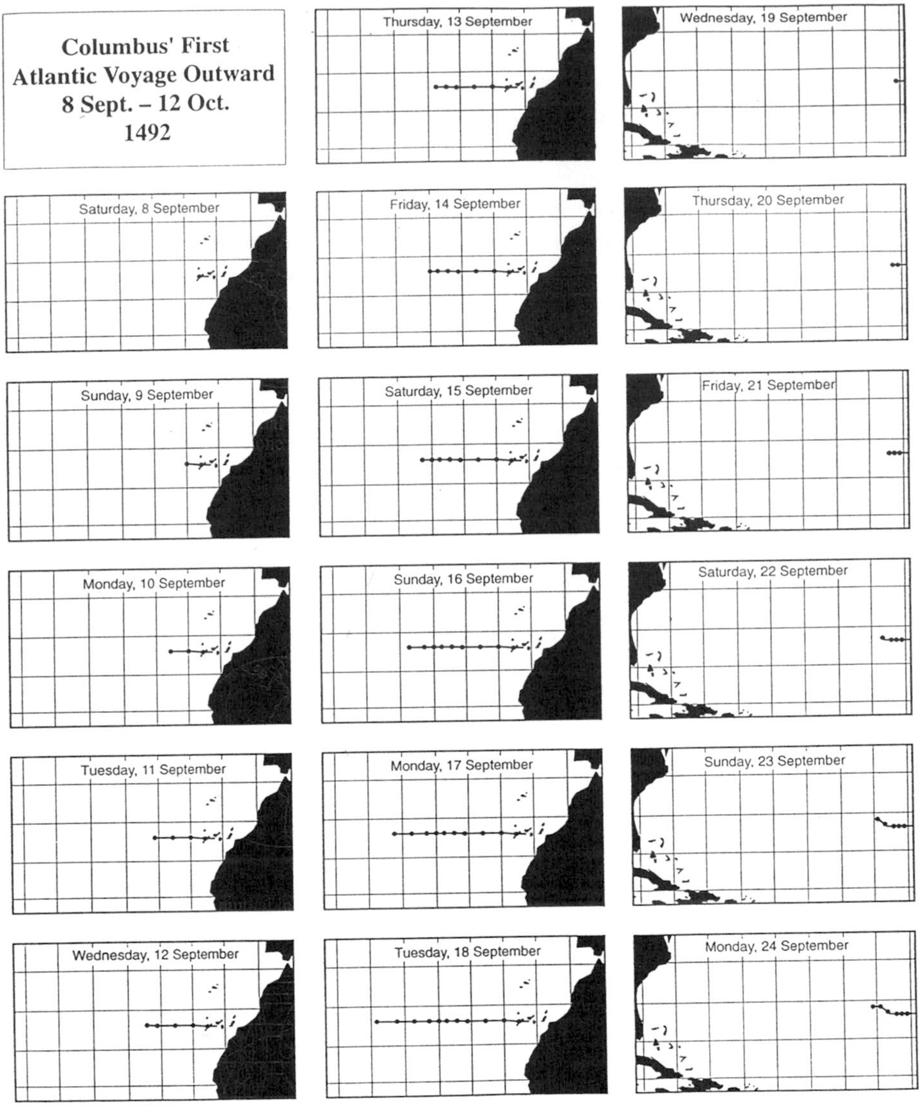

Thursday, 13 September

Wednesday, 19 September

Saturday, 8 September

Friday, 14 September

Thursday, 20 September

Sunday, 9 September

Saturday, 15 September

Friday, 21 September

Monday, 10 September

Sunday, 16 September

Saturday, 22 September

Tuesday, 11 September

Monday, 17 September

Sunday, 23 September

Wednesday, 12 September

Tuesday, 18 September

Monday, 24 September

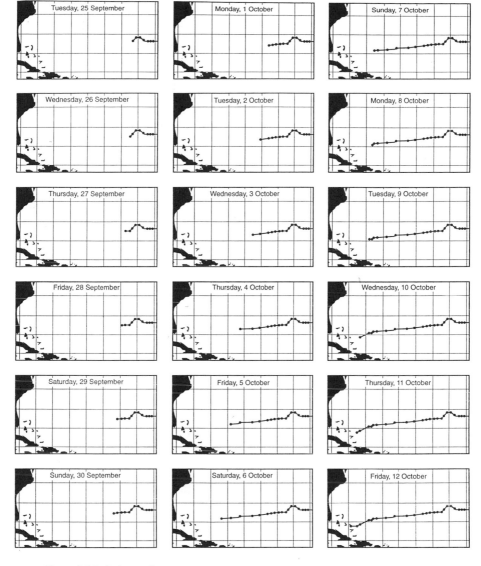

*Figure 9.14.* Animated map sequence showing the track of Christopher Columbus on his first voyage to the New World in 1492. Each day's journey is marked with a dot and a line; the exposure time for every frame is the same making the time (temporal) scale correct.

its own methodology and traditions. Sometimes these have had an inhib-
iting effect on progress, but in spite of a tendency toward conservatism,
cartographers have generally been sensitive to change and kept abreast
of philosophical, artistic, scientific, and technological progress. Geographically, the centers of action in cartography have usually been where
science is currently flourishing. As in the past, ingenious solutions will
undoubtedly be found to problems associated with the representation
of spatial phenomena in the future, while "knowledge," as Tennyson re-
minds us, continues to "grow from more to more."

## *Appendix A*  Selected Map Projections

| Name of Projection | Century | Inventor | Family/Form | Salient Characteristics | Principal Uses |
|---|---|---|---|---|---|
| Gnomonic (horoscope) | 5th B.C. | Thales? | Azimuthal | All straight lines are great circles; limited area coverage | Astronomy; later, route plotting |
| Orthographic (analemma) | 2d B.C. | Hipparchus? | Azimuthal | A hemisphere as viewed from infinity | Astronomy; later, earth illustrations |
| Stereographic (planisphere) | 2d B.C. | Hipparchus? | Azimuthal | Conformal; circles project as circles | Astronomy; later, air navigation |
| Marinus (plane chart/plate carée) | 1st A.D. | Marinus of Tyre | Cylindrical | Straight, equally spaced meridians and parallels (all meridians and the equator true to scale) | Early maps of known world/regions |
| Ptolemy (first projection) | 2d | C. Ptolemy | Coniclike (equidistant from origin) | Straight, radiating meridians; curved, concentric parallels including, theoretically, the N. Pole | Early maps of known world/regions |

| Name of Projection | Century | Inventor | Family/Form | Salient Characteristics | Principal Uses |
|---|---|---|---|---|---|
| Ptolemy (second projection) | 2d | C. Ptolemy | Pseudoconic | Curved meridians; curved parallels | Early maps of known world/regions |
| Contarini | 16th | G. Contarini | Coniclike (equidistant from curved N. Pole) | Straight, radiating meridians; curved, concentric parallels, including N. Pole | Early maps of Old and New Worlds |
| Ruysch | 16th | J. Ruysch | Coniclike (equidistant from N. Pole) | Straight, radiating meridians; curved, concentric parallels; N. Pole at a point | Early maps of Old and New Worlds |
| Waldseemüller | 16th | M. Waldseemüller | Miscellaneous | Curved and "bent" meridians; curved parallels | Early maps of Old and New Worlds |
| Rosselli | 16th | F. Rosselli | Oval | Curved, elliptical meridians; straight, equally spaced parallels; central hemisphere is a circle | Early maps of Old and New Worlds |
| Maggioli | 16th | V. de Maggioli | Azimuthal (equidistant from N. Pole) | Straight, radiating meridians; curved, concentric parallels equidistant from N. Pole to equator | Polar regions (part) to 35° S (part) |

| | | | | | |
|---|---|---|---|---|---|
| Vespucci | 16th | J. Vespucci | Azimuthal (equidistant from N. and S. Poles) | Straight, radiating meridians; curved, concentric parallels; interrupted | World in one hemisphere and two quarter spheres |
| Agnese | 16th | B. Agnese | Ovoid (pseudoconic) | Curved meridians; straight, equally spaced parallels; poles half the length of equator | World maps |
| Cordiform (Werner) | 16th | B. Sylvanus J. Stabius J. Werner O. Fine et al. | Miscellaneous | Curved meridians; curved concentric parallels; equal-area in some cases | World maps |
| Double cordiform | 16th | O. Fine G. Mercator | Miscellaneous | Curved meridians; curved, concentric parallels; equal-area; interrupted | World maps |
| Conic (with two standard parallels) | 16th | G. Mercator | Conic | Less deformation than simple conic | Continents, countries, regions |

| Name of Projection | Century | Inventor | Family/Form | Salient Characteristics | Principal Uses |
|---|---|---|---|---|---|
| Mercator (nautical or mariner's) | 16th | G. Mercator | Cylindrical | Straight, equally spaced meridians; straight parallels; spacing increases with increasing degrees from equator; straight lines, loxodromes; conformal | Navigational charts; world maps |
| Sinusoidal (Sanson–Flamsteed) | 16th? | N. Sanson and J. Flamsteed used this projection (which had been invented earlier) in the 17th century and are sometimes credited with it | Miscellaneous | Equivalent; meridians are sine curves; parallels are straight lines, equally spaced | World distributions; atlas maps |
| La Hire | 18th | P. de La Hire | Azimuthal | Perspective; straight meridians radiating from the pole; curved concentric parallels with 45° half the circumference of the equator | Hemispheric maps |

| | | | | | |
|---|---|---|---|---|---|
| Cassini (Soldner) | 18th | C.-F. Cassini | Cylindrical | Transverse; straight central meridian and lines perpendicular are true to scale | Early topographic maps |
| Bonne | 18th? | R. Bonne in dependence on others (Ptolemy, Werner et al.) | Miscellaneous | Equivalent; central straight meridian true to scale, as are all curved, concentric parallels | Topographic maps, esp. of midlatitudes |
| Murdoch | 18th | P. Murdoch | Conic | Equidistant; straight, radiating meridians; curved, concentric parallels | Midlatitude maps |
| Lagrange | 18th | J. H. Lambert, promoted by J. Lagrange | Circular | Conformal, except at poles; all meridians and parallels are circular arcs | World maps |
| Transverse cylindrical equal-area | 18th | J. H. Lambert | Cylindrical | Equivalent; straight central meridian true to scale, others curved; straight equator, other parallels curved | Topographic maps |

| Name of Projection | Century | Inventor | Family/Form | Salient Characteristics | Principal Uses |
|---|---|---|---|---|---|
| Conic equal-area | 18th | J. H. Lambert | Conic | Equivalent; straight meridians radiate from pole; concentric parallels (one standard) with spacing decreasing from poles | Distributional maps, esp. of midlatitudes |
| Conic conformal | 18th | J. H. Lambert | Conic | True shape around a point; radiating meridians; two standard parallels | Midlatitude maps; air navigation |
| Cylindrical equal-area | 18th | J. H. Lambert | Cylindrical | Straight, equally spaced meridians; spacing of straight parallels decreases from equator to poles | Distributional maps of the world or parts |
| Polar azimuthal equal-area | 18th | J. H. Lambert | Azimuthal | Equivalent; straight, radiating meridians, directions true from a point (pole); curved, concentric parallels | Hemispheres, polar areas, later, airline maps |
| Equatorial azimuthal equal-area | 18th | J. H. Lambert | Azimuthal | Equivalent; straight central meridian, others curved and spaced closer toward edge; straight equator, other parallels curved | Hemispheres, atlas maps |

| Name | Century | Inventor | Type | Properties | Uses |
|---|---|---|---|---|---|
| Transverse Mercator | 18th | J. H. Lambert; often attributed to C. F. Gauss | Cylindrical | Conformal; straight central meridian true to scale; equator straight, other parallels and meridians curved | Topographic maps; later, airline routes |
| Albers | 19th | H. C. Albers | Conic | Equivalent; straight, radiating meridians; curved parallels, two standard, others decreasing in spacing away from these | Topographic maps; continents, especially in the midlatitudes |
| Mollweide (homolographic) | 19th | K. B. Mollweide | Oval | Equivalent; curved, elliptical meridians; straight parallels with spacing decreasing from equator to poles; central hemisphere is a circle (see Rosselli, above) | World distributions; atlas maps |
| Polyconic (ordinary polyconic) | 19th | F. Hassler | Miscellaneous (Conic-type) | Curved meridians; curved parallels along which scale is correct; adapted in the 20th century for IMW maps | Topographic maps and coastal charts; with modification, world map series |

| Name of Projection | Century | Inventor | Family/Form | Salient Characteristics | Principal Uses |
|---|---|---|---|---|---|
| Ellipsoidal transverse Mercator | 19th | C. F. Gauss | Cylindrical | Conformal; straight central meridian, correct in scale; curved parallels, equally spaced on central meridian | Topographic maps, esp. for areas with a greater N–S than E–W extent |
| Gall orthographic | 19th | J. Gall (sometimes attributed to A. Peters; see text for Gall's two other cylindrical projections) | Cylindrical | Equivalent; straight, equally spaced meridians; straight parallels, 45° N and 45° S are standard | World distribution maps |
| Van der Grinten I | 20th | A. J. Van der Grinten (see text for Van der Grinten's three other circular projections) | Circular | Meridians, circular arcs (central meridian, straight line); parallels circular arcs (equator, straight line) | Wall maps; illustrative maps in journals and books |

| | | | | | |
|---|---|---|---|---|---|
| Eckert IV | 20th | M. Eckert (later, Eckert-Griefendorff; see text for his five other pseudocylindrical projections) | Pseudocylindrical | Equivalent; meridians semi-ellipses (central meridian, straight line); parallels straight lines, spacing decreasing toward poles; based on two circles (see Agnese, above) | Atlases; distributional maps |
| Goode homolosine | 20th | J. P. Goode (also interrupted and compressed projections) | Miscellaneous | Equivalent; poleward part from Mollweide (see above) and equatorial part from sinusoidal (see above) | Atlases; distributional maps |
| Miller (modified Mercator) | 20th | O. M. Miller (also proposed three other cylindrical projections) | Cylindrical | Straight meridians; straight parallels (30° N and S standard); allows the poles to be shown, a compromise projection | Atlas maps |
| Fuller Dymaxion | 20th | R. B. Fuller | Miscellaneous | Polyhedrons; eight equilateral triangles and six squares, or twenty equilateral triangles with constant scale along edges | Models simulating the globe or illustrating global relationships |

| Name of Projection | Century | Inventor | Family/Form | Salient Characteristics | Principal Uses |
|---|---|---|---|---|---|
| Robinson orthophanic | 20th | A. H. Robinson | Pseudocylindrical | Curved meridians (central meridian, straight line); straight parallels; spacing to produce a compromise between conformality and equivalence | Atlas maps; maps in journals |
| Perspective orthographic | ? | Hipparchus (see above, but used recently by F. Debenham, R. E. Harrison, et al. and for GOES satellite images) | Azimuthal | Hemisphere as viewed from "infinity"; meridians and parallels in perspective | Atlas maps; maps in journals; television weather maps |

# Appendix B

## Short List of Isograms

*Isogram, isarithm,* and *isoline* are generic terms embracing both isometric lines and isopleths. An isometric line is a line representing a constant value obtained from measurement at a series of points, such as a contour line. An isopleth is a line connecting points assumed to have equal value, such as an isodem.[1] *Isogram,* proposed in 1889 by Sir Francis Galton, is the most general of these terms.

The following are selected from a much larger number of terms for special forms of the isogram. They are presented in chronological order according to their assumed first cartographic use. The contemporary terms, as listed, were not in every instance applied for their earliest cartographic usage. (For example, Halley used the term *curve lines* for *isogones,* and they were known as *Halleyan* or *Halleian lines* for about a century before receiving their present designation from the Norwegian astronomer Christopher Hansteen around 1820.) Each definition should be preceded by the words "line along which the . . ." and followed by ". . . is, or is assumed to be, the same or constant."[2]

| | |
|---|---|
| Isobath | depth below a datum (for example, mean sea level) |
| Isogonic line | magnetic declination |
| Isocline | magnetic dip (inclination) or angle of slope |
| Isohypse (contour) | elevation above a datum (for example, mean sea level) |
| Isodynamic line | value of intensity or a component of the intensity of the magnetic field |
| Isotherm | temperature (usually average) |
| Isobar | atmospheric pressure (usually average) |
| Isohyet | precipitation |
| Isobront | occurrence of thunderstorms |
| Isanther | time of flowering of plants |
| Isopag | duration of ice cover |
| Isodem | population |
| Isoamplitude | amplitude of variation (often of annual temperature) |
| Isoseismal line | number (or intensity) of earthquake tremors |
| Isochasm | annual frequency of aurorae |

| Isophot | intensity of light on a surface |
| Isoneph | degree of cloudiness (often average) |
| Isochrone | travel time from a given point |
| Isophene | date of beginning of a plant species entering a certain phenological phase |
| Isopectic | time of ice formation |
| Isotac | time of thawing |
| Isobase | vertical earth movement |
| Isohemeric line | minimum time of (freight) transportation |
| Isohel | average duration of sunshine in a specified time |
| Isodopane | cost of travel time |
| Isotim | the price for certain goods |
| Isoanabase | rising of land in relation to a coast |
| Isophort | freight rates on land |
| Isonau | freight rates at sea |
| Isomist | wages |
| Isothym | intensity of evaporation |
| Isoceph | cranial indices |
| Isohalaz | frequency of hailstorms |
| Iosgene | density of a genus |
| Isospecie | density of a species |
| Isodyn | economic attraction |
| Isohydrodynam | potential water power |
| Isostalak | intensity of plankton precipitation |
| Isovapor | vapor content in the air |
| Isodynam | traffic tension |
| Isohygrom | number of arid or humid months per year |
| Isobenth | amounts of zoobenthos per unit area at given depths |
| Isonoet | equal average degree of intelligence |
| Isopach | thickness of sedimentary deposits |

# Appendix C

## Glossary

The list of predominantly contemporary cartographic terms that follows was compiled mainly from the text. Foreign words and proper names were excluded, as well as any except the most general projection and isoline terms, which are the subjects of appendixes A and B, respectively. The list was then submitted to a group of geography students specializing in cartography. If a majority of them thought a term was already too well known to require definition, it was eliminated from the list. The remaining terms were then examined in the text to see if they were adequately defined there; if so, they were eliminated. The meanings of those still remaining were checked in a standard college dictionary, such as any student might possess, and if they were suitably defined for the purpose of this work, they were dropped from the list. Hence, the glossary contains only a selection of the technical terms, including specialized uses of some common words, in this book. A number of sources were consulted in its compilation. However, many definitions are adapted and a few taken directly from the *Glossary of Mapping, Charting, and Geodetic Terms,* 2d ed. (Washington, D.C.: Department of Defense, Department of the Army, Corps of Engineers, U.S. Army Topographic Command, 1969); some are derived from the *Glossary of Technical Terms in Cartography* (London: The Royal Society, 1966) and others from Robert N. Colwell, ed., *Manual of Remote Sensing,* 2d ed. (Falls Church, Va.: American Society of Photogrammetry, 1983), 1:1183–98. These compilations contain a very large number of definitions and should be consulted for additional terms not included in the following list.

**aerial survey**  Mapping utilizing photographic, electronic, or other data obtained from an airborne station. Also called *air survey.*

**altazimuthal theodolite**  An instrument equipped with both horizontal and vertical graduated circles for the simultaneous observation of horizontal and vertical directions or angles.

**altitude tinting**  *See* hypsometric tinting.

**anaglyph**  A stereogram in which the two views are printed or projected superimposed in complementary colors, usually red and blue. When viewed through filter spectacles of corresponding colors, a stereoscopic image is formed.

249

**arc of the meridian**  A part of an astronomic or geodetic line of longitude.

**area proportional to (APT) map**  A cartogram in which the surface extent of features is relative to the amount of the map data (such as population) rather than the geographical extent of the base to which it is related. Also called *area cartogram.*

**area[1] symbol**  A continuous and distinctive shading, tone, or repetitive pattern used on a map to represent features, real or theoretical, usually with considerable surface extent (forests, religion). It contrasts with a point symbol or line symbol (q.v.).

**area reference grid**  A plane-rectangular coordinate system usually based on and mathematically adjusted to a map projection, with numbers and/or letters used to designate positions of reference to the system.

**astronomical north**  *See* north.

**azimuth**  The horizontal direction of a line measured clockwise from a reference plane, usually the meridian.

**azimuthal map projection**  A systematic representation of the graticule on which the directions of all lines radiating from a central point or pole are the same as the directions of the corresponding lines on the graticule.

**bar scale**  *See* graphical scale.

**base data**  Fundamental cartographic information (such as coastlines, political boundaries) in relation to which additional data of a more specialized nature may be compiled or overprinted.

**base line**  A surveyed line established with more than usual care, to which surveys are referred for coordination or correlation.

**block diagram**  A representation of a landscape usually in perspective or isometric projection, frequently exaggerated in the vertical scale.

**cadastral map**  A plan showing the boundaries of subdivisions of land, usually with bearings and lengths and the areas of individual tracts, for purposes of describing and recording ownership.

**cardinal direction**  Any of the four principal astronomical directions on the surface of the earth: north, east, south, or west.

**cartobibliography**  A systematic list of maps, usually relating to a given region, subject, or person.

**cartogram**  An abstracted or simplified map for displaying quantitative data for which the base is normally not true to scale.

**cartography**  The production—including design, compilation, construction, projection, reproduction, use, and distribution—of maps.

**cartouche**  A feature of a map or chart, often a decorative inset, containing the title, legend, scale, or all of these items.

**central meridian**  A great circle of the earth's system of longitude used in the construction of map projections, especially pseudocylindrical projections; several might be used in the case of an interrupted projection.

**chorographic-scale map**  A systematic representation of an intermediate-sized land area (a country), in contrast to a small-scale map or a large-scale map (q.v.). Also called an *intermediate-scale map.*

**choropleth map**   A systematic representation in which color or shading is applied to areas bounded by statistical or administrative limits.

**circle of illumination**   The great circle that is the edge of the sunlit hemisphere, dividing the earth between a light half and a dark half at any given moment.

**color infrared (CIR)**   A photographic film that senses in the near-infrared portion of the electromagnetic spectrum and filters out the blue that produces false color images on which (for example) live vegetation appears red. Not to be confused with thermal infrared (TIR).

**color separation**   The process of preparing a separate drawing, engraving, or negative for each color required in the production of a lithographed map or chart.

**compass north**   *See* north.

**compass rose**   A circle graduated from the reference direction, usually north, in compass points, degrees (0–360), or both.

**condensed projection**   A systematic representation of the graticule with areas of little or no importance for a particular purpose eliminated and the remainder brought close together.

**conformal map projection**   A systematic representation of the graticule on which the shape of any small area of the surface mapped is unchanged; also called an *orthomorphic map projection,* it contrasts with an equal-area map projection (q.v.).

**contour**   An imaginary line connecting all points that are at the same elevation above or below a datum surface, usually mean sea level (an isohypse).

**contour interval**   The vertical distance between two adjacent contour lines.

**controlled mosaic**   An assemblage, usually of rectified aerial photographs, oriented and scaled to horizontal ground control.

**coordinate system**   A graticule, or a Cartesian grid, in which points are located from two (or three) axes that intersect at a point.

**cosmography**   The description and mapping of the heavens and the earth, including astronomy, geography, and geology.

**cosmology**   The study of the origin and structure of the universe, including elements, laws, space, and time.

**dasymetric map**   A representation in which color or shading is applied to areas that have homogeneity, within specified limits, and in which it is not necessary for the color or shading to be limited by statistical or administrative boundaries.

**datum**   Any numerical or geometrical value, surface, line, or point that may serve as a base or reference for other quantities.

**dead reckoning**   Calculation of position (usually at sea) on the basis of distance traveled and time taken, with allowance made for winds, currents, and so forth in the case of sailing.

**declination**   *See* magnetic declination.

**deformation**   *See* map distortion.

**density symbol**   Shading or color used to cartographically represent quantity; usually, the greater the amount, the deeper the shading or color.

**depression contour**   A line joining points of equal value around a closed feature such as a crater, usually indicated with short lines perpendicular to the contour line in the direction of lower elevation.

**dimensional stability**　The ability of material to maintain size caused by changes in moisture and temperature.

**distortion**　*See* map distortion.

**dot map**　A systematic representation of earth phenomena in which dots (usually of uniform size) each represent a specific number of the distribution being mapped.

**electromagnetic spectrum (EMS)**　The ordered array of known electromagnetic radiations, from short (cosmic) to long (radio) and including visible rays.

**equal-area map projection**　A systematic representation of the graticule on which the area of any enclosed figure on the map is equal to the area of the corresponding figure on a globe of the same scale; contrasts with a conformal projection (q.v.). Also called *equivalent map projection*.

**field survey**　*See* ground survey.

**flow line**　A linear cartographic symbol in which width varies in proportion to the quantity being mapped.

**form line**　A linear symbol resembling a contour, but often broken or dashed, representing only approximate elevation and used to show the shape of the terrain rather than actual height.

**four-color process**　*See* process color.

**fractional scale**　*See* representative fraction.

**general map**　A systematic representation of an area showing a variety of geographical phenomena (coastlines, political boundaries, transportation lines) used for planning, location, reference, and so forth; contrasts with a thematic map (q.v.).

**generating globe**　A model of the sphere used for the development of perspective map projections, or a theoretical sphere to which projections may be referred for comparative purposes. The radius of the generating globe bears the same relationship to the earth as is denoted by the representative fraction of the resulting map.

**geocartography**　The mapping of earth phenomena, in contrast to the mapping of extraterrestrial and other bodies.

**geodesy**　The study that deals with the measurement, shape, and size of the earth and of large areas of the globe, such as countries.

**geographical north**　*See* north.

**geographical-scale map**　*See* small-scale map.

**geographic information systems (GIS)**　An electronic method of acquiring, processing, storing, managing, reproducing, displaying, and analyzing spatial data; subsumed under GIS is conventional cartography and remote sensing of the environment.

**globe gore**　A lune-shaped segment that can be fitted to the surface of a sphere with little distortion or deformation.

**graduated circle** A disc-shaped symbol proportional in actual area or appearance to the amount of the phenomenon being mapped relative to other similarly shaped symbols. Also called a *proportional circle.*

**graphical scale** A graduated line by means of which distances on a map or chart may be measured in terms of ground distances; also known as a *bar scale* or *linear scale.*

**graticule** A network of lines representing the earth's parallels of latitude and meridians of longitude.

**great circle** A line on the earth's surface, the plane of which passes through the center of the globe. This shortest distance between two points on the sphere is also known as an *orthodrome.*

**grid** A Cartesian reference system consisting of two sets of parallel lines intersecting at right angles, forming squares; also used loosely in reference to the (earth) graticule (q.v.).

**ground survey** Measurement and mapping in the field, as distinguished from aerial survey (q.v.).

**hachure** A short line running in the direction of maximum slope to indicate the relief of the land in relation to other such lines by thickness and spacing.

**halftone** A process by which gradations between black and white are obtained by a system of dots produced through a screen placed between a camera and a sensitized plate.

**high latitude** A polar or subpolar area of the earth.

**hypsometric tinting** A method of showing relief on maps and charts by coloring, in different shades, those parts that lie between different levels (elevations).

**image** The recorded representation and counterpart of an object, landscape, or other item as a map, chart, plan, photograph, or radar scan.

**inset map** A separate map positioned within the borders of a larger map and usually of a different scale.

**intermediate-scale map** *See* chorographic-scale map.

**interrupted map projection** A systematic representation of a graticule in which the origin or central meridian is repeated in order to reduce peripheral distortion; also known as *recentered projection.*

**inverted image** *See* pseudoscopic image.

**isarithm, isogram, isoline** *See appendix B.*

**isometric diagram** A representation simulating the third dimension, in which the scale is correct along three axes.

**large-scale map** A systematic representation of a small land area (with a representative fraction arbitrarily set at 1:75,000 or greater); also sometimes called a *topographic-scale map.*

**layer tinting** *See* hypsometric tinting.

**legend** An explanation of, or key to, the cartographic symbols used on a map, diagram, or model.

**leveling**   The operation of measuring vertical distances, directly or indirectly, to determine elevations.

**libration**   A real or apparent oscillatory motion, particularly on the moon. This results in more than half of the moon's surface being revealed to an observer on the earth even though the same side of the moon is always toward the earth.

**linear scale**   *See* graphical scale.

**line[ar] symbol**   A distinctive line used to represent features, real or theoretical, that have length but little or no width (roads, political boundaries).

**low latitude**   A tropical or subtropical area of the earth.

**loxodrome**   *See* rhumb line.

**magnetic declination**   The angle between the magnetic and geographic meridians at any place, expressed in degrees east or west to indicate the direction of magnetic north from true north.

**magnetic north**   *See* north.

**magnetic variation**   A term used as a synonym for *magnetic declination* but, more specifically, to indicate changes in this relationship within certain time limits (temporal variation).

**map**   A representation, usually on a plane surface, of all or part of the earth or some other body showing a group of features in terms of their relative size and position.

**map data**   Specific cartographic information plotted in relation to base data (q.v.).

**map distortion**   Alteration in the shape of a cartographic representation caused by the transformation of the sphere or spheroid (or part of such a figure) through projection onto a plane surface; also called *map deformation.*

**map projection**   Any systematic arrangement of the meridians and parallels (graticule) of the all-side-curving figure of a sphere or spheroid (or a part of such a figure) on a plane surface.

**mean sea level (MSL)**   The average height of the surface of the sea over all stages of the tide.

**metes and bounds survey**   The description of the boundaries of tracts of land (such as properties) based on the bearing and length of each successive line, often keyed to an ownership list.

**midlatitude**   An area between the subtropical and subpolar areas of the earth.

**mosaic**   *See* controlled mosaic *and* uncontrolled mosaic.

**natural scale**   *See* representative fraction.

**normal case of a projection**   The mathematically simplest aspect of a representation of the graticule (typically, principal directions in the representation coincide with those on the graticule). *See* oblique map projection *and* transverse map projection.

**north**   The primary reference direction relative to earth. Magnetic or compass north is the direction of the north-seeking end of a magnetic compass needle not subject to local disturbance. True, astronomical, or geographical north is the northern direction of the meridian at the point of observation. Grid north is the direction of the north–south lines on a map, coincident with true north only at the meridian of origin.

**oblate spheroid**    An ellipsoid of rotation, the shorter axis of which is the axis of rotation. The earth is approximately an oblate spheroid.

**oblique map projection**    A systematic representation of the graticule with an axis inclined at an angle between 0 and 90 degrees but equal to neither. *See* normal case of a projection *and* transverse map projection.

**orientation**    The act of establishing, or the state of being in, correct relationship in direction with reference to the points of the compass.

**orthodrome**    *See* great circle.

**orthomorphic map projection**    *See* conformal map projection.

**parallax**    The apparent displacement of the position of a body with respect to a reference point or system, caused by a shift in the point of observation.

**perspective diagram**    A representation simulating the third dimension, with the appearance to the eye of objects correct in respect to their relative distance and position.

**photogrammetry**    The science or art of obtaining reliable measurements and/ or preparing maps and charts from aerial photographs using stereoscopic equipment and methods.

**photolithography**    A method of printing in which the original subject is photographed and the consequent image is transferred to a plate (usually grained metal) for lithographic printing. Loosely used to refer to the whole process of *lithography,* and vice versa.

**pie graph**    Circular symbol divided into sectors to indicate proportions of a total value. Also known as a *sectored circle.*

**planimetric map**    A systematic representation of land with only the horizontal positions of features shown. Contrasts with a topographic map (q.v.).

**plastic scribing**    *See* scribing.

**plastic shading**    *See* shaded relief.

**point symbol**    A distinctive device used to represent features, real or theoretical, usually with limited areal extent (settlements). However, such symbols are sometimes used in combination with other symbols to show density, as on a dot map (q.v.).

**poles**    The extremities of the axis of rotation of the earth (North and South Poles); also two points on the earth's surface where the needle of the magnetic compass stands vertical (North and South Magnetic Poles, respectively relatively close to, but not coincident with, the North and South Poles as defined above).

**prime meridian**    The north–south line on the earth from which longitude is measured. Over the centuries, several prime meridians and numbering systems have been used; today, by international agreement, longitude is usually reckoned east 180 degrees and west 180 degrees of Greenwich, England (0 degrees longitude).

**process color** A photomechanical method of printing in which the separation of the colors of the original is accomplished mechanically and photographically. It includes, as a special case, four-color process, in which filters and screens are used to break images into four colors (yellow, magenta, cyan, and black) that, when recombined at the printing stage, will simulate essentially all colors in the original. For video the colors are red, green, and blue.

**profile** A vertical cross section of the surface of the earth and/or the underlying strata along any fixed line. It often involves vertical exaggeration (q.v.).

**prolate spheroid** An ellipsoid of rotation, the longer axis of which is the axis of rotation.

**proportional circle** *See* graduated circle.

**pseudocylindrical projection** An arrangement of the earth grid in which, typically, parallels are straight lines that vary in length, with curved meridians equally spaced on the parallels. Some pseudocylindrical projections are equivalent (Eckert IV), but none are conformal.

**pseudoscopic image** A three-dimensional impression that is the reverse of that actually existing (as in photographs, shading, and so forth). Also called an *inverted image.*

**quadrangle (quad.)** A single sheet of a standard topographic map series, such as those of the United States Geological Survey.

**range line** In the United States Public Land Survey, a boundary of a township (q.v.), surveyed in a north–south direction.

**recentered map projection** *See* interrupted map projection.

**reconnaissance map** The cartographic product of a preliminary examination or survey of an area and therefore of a lower order of accuracy than later, more rigorous surveys.

**remote sensing** The detection and/or recording of data about an object in which the sensor is not in direct physical contact with the object.

**representative fraction (RF)** The scale of a map or chart expressed as a fraction or ratio that relates unit distance on the map to distance measured in the same unit on the ground (1:1,000,000). Also called a *natural scale* or *fractional scale.*

**resolution, spatial** The ability of a system (remote sensing) to render a sharply defined image, which may be expressed as lines per millimeter, temperature, or other physical property.

**rhumb line** A line on the surface of the earth making the same angle with all meridians. Also called a *loxodrome* or *line of constant compass bearing*, it spirals toward the poles in a constant, true direction.

**riparian survey** The mapping of areas along the banks of a river; early reconnaissance surveys often followed the routes of explorers along major streams such as the Nile and the Mississippi.

**scale** The ratio of a distance on a map, globe, model, or photograph to its corresponding distance on the ground or on another graphical representation.

**scan**   A narrow strip of the earth that is swept by any device (such as radar), producing a series of lines that together make up an image (q.v.); this stands in contrast to a snapshot or normal photograph, in which the whole scene is imaged simultaneously.

**scribing**   The process of preparing a negative (or positive) that can be reproduced by contact exposure. Portions of a photographically opaque coating are removed from a transparent (usually plastic) base with specially designed tools.

**section**   In the United States Public Land Survey, the unit of subdivision of a township, normally a quadrangle of one square mile. There are thirty-six such units in a township (q.v.).

**sectored circle**   *See* pie graph.

**shaded relief**   The rendering of landforms by continuous graded tone to give the appearance of shadows thrown by a light source, normally located above the northwest of the map.

**small-scale map**   A systematic representation of a large land area; also called a *geographical-scale map.*

**spherical coordinates**   A system of polar coordinates in which the origin is in the center of the sphere and the points all lie on the surface. Also loosely known as a *spherical grid.*

**spheroid**   Any figure differing slightly from a sphere; in geodesy, one of several mathematical figures closely approaching the undisturbed mean sea level of the earth extending continuously through the continents (geoid), used as a surface of reference for geodetic surveys.

**spot elevation**   A point on a map or chart, usually marked by a dot, with a numerical expression of elevation; also called *spot height.*

**standard line**   A parallel, meridian, or other basic linear feature of a map projection, along which the scale is as stated on the map or chart and which is used as a control line in the computation of a map projection. Also called *standard meridian* or *standard parallel.*

**statistical surface**   A theoretical three-dimensional figure resulting from isopleth, choropleth, or other forms of quantitative mapping.

**stereoscope**   A binocular optical instrument to assist an observer in viewing photographs and diagrams to obtain a mental impression of a three-dimensional model.

**strip map**   A cartographic device for showing, in diagrammatic form, routes from one point to another along a more or less straight line.

**symbol**   A diagram, design, letter, character, or abbreviation placed on maps, charts, and other models that by convention, usage, or reference to a legend is understood to stand for, or represent, a specific characteristic or feature. It may be in the form of an areal, linear, point, or other symbol (q.v.).

**synoptic chart**   A systematic representation to indicate conditions prevailing or predicted to prevail over a considerable area at a given time (for example, a weather map).

**thematic map**   A systematic representation of an area normally featuring a single distribution as its map data (for example, population) and for which the base data serve only to help locate the distribution being mapped. In its function it contrasts with a general map (q.v.).

**topographic map**   A systematic representation of a small part of the land surface showing physical features (relief, hydrography) and cultural features (roads, administrative boundaries). These large-scale maps present both vertical and horizontal features in measurable form.

**topographic-scale map**   *See* large-scale map.

**toponym**   A place-name, or a word derived from a geographical place.

**township**   In the United States Public Land Survey, a quadrangle of approximately six miles on each side, consisting of thirty-six sections.

**township line**   In the United States Public Land Survey, a boundary of a township (q.v.) surveyed in an east–west direction. See also *range line*.

**transverse map projection**   A systematic representation of the graticule with its axis rotated 90 degrees (right angle) to that considered the normal case of a map projection (q.v.) in any particular example. *See also* oblique map projection.

**uncontrolled mosaic**   An assemblage of unrectified prints, the detail of which has been matched from print to print without ground control or other orientation.

**variation**   *See* magnetic variation.

**verbal scale**   An expression of the relationship between specific units of measure on the map and distance on the ground ("one inch equals one mile"); a less general expression than the representative fraction (q.v.) (in this case, 1:63,360).

**vertical exaggeration**   The change in a model surface or profile created by proportionally raising the apparent height of all points above the base level while retaining the same base.

**volumetric symbol**   A cartographic device (simulated sphere) to give a quantitative impression of the third dimension.

**zenithal map projection**   *See* azimuthal map projection.

# Notes

## Chapter One

1. H. Marshall McLuhan, *Understanding Media* (New York: McGraw-Hill, 1964), esp. 157–58. This book, and other writings by this author, have much relevance to cartography as a means of communication, even if we do not accept the full implications of the phrase "the medium is the message."

2. General textbooks on modern (thematic) mapmaking techniques that are well known to American geographers include Erwin J. Raisz, *General Cartography*, 2d ed. (New York: McGraw-Hill, 1948); Arthur H. Robinson and others, *Elements of Cartography*, 5th ed. (New York: John Wiley, 1984); F. J. Monkhouse and H. R. Wilkinson, *Maps and Diagrams*, 2d ed. (London: Methuen, 1963); John Campbell, *Introductory Cartography* (Englewood Cliffs, N.J.: Prentice-Hall, 1984); Borden Dent, *Principles of Thematic Map Design* (Reading, Mass.: Addison-Wesley, 1985); Judith Tyner, *Introduction to Thematic Cartography* (Englewood Cliffs, N.J.: Prentice-Hall, 1992), and R. W. Anson and F. J. Ormeling, *Basic Cartography for Students and Technicians*, 2d ed., vol. 1 (Oxford: Elsevier, 1994). Map-reading works beyond the elementary level include: *United States Department of the Army Field Manual 21–26* (Washington, D.C.: Government Printing Office, 1965); Judith Tyner, *The World of Maps and Mapping* (New York: McGraw-Hill, 1973); Mark Monmonier and George A. Schnell, *Map Appreciation* (Englewood Cliffs, N.J.: Prentice-Hall, 1987); and Philip C. Muehrcke and Juliana O. Muehrcke, *Map Use: Reading, Analysis, Interpretation*, 3d ed. (Madison: J. P. Publications, 1992).

3. Many countries maintain a map library as part of their national collection. In the United States, the most comprehensive assemblage of maps and atlases—early and modern, foreign and domestic—is contained in the Geography and Map Division of the Library of Congress, while the National Archives is the official depository of U.S. government maps. The Map Division of the British Library, London, and the Bibliothèque Nationale, Paris, have exceptionally rich and varied cartographic holdings. Big city libraries often have large map collections, as do some government agencies and geographical and other scientific organizations. Larger universities often have important map resources, although these are frequently scattered among different departments. However, in some cases they are centralized, as at the University of California, Los Angeles, where

over half a million sheets published since 1900 as well as aerial photographs are contained in the UCLA Map Library, and maps and atlases published before this date are stored in Special Collections. See Walter W. Ristow, "The Emergence of Maps in Libraries," *Special Libraries* 58, no. 6 (July–August 1967): 400–419. The Map and Geography section of the Special Libraries Association has done much to promote cartography. An international center with the world's cartographic records on film or tape has been suggested but has not yet been realized. Government agencies, especially those concerned with defense in various countries, have large collections of maps and charts, but these are not normally accessible to the general public.

4. A number of periodicals, especially geography journals, occasionally include articles on map topics. In addition, there are several national and international journals concerned exclusively with cartography. Among the latter group are *The International Yearbook of Cartography* (Gütersloh: Bertelsmann Verlag), founded by Eduard Imhof in 1961 and published annually since that date as the organ of the International Cartographic Association; and *Imago Mundi: A Periodical Review of Early Cartography* which was founded by Leo Bagrow in 1935 (the first annual number being published in Berlin but subsequent issues in England; a second series began in 1975 with numbering continued from the original series). In 1994 it was subtitled *The International Journal for the History of Cartography*. See also Chauncy D. Harris and Jerome D. Fellman, *International List of Geographical Serials,* University of Chicago, Department of Geography Research Paper 63 (Chicago, 1960), and Chauncy D. Harris, *Annotated World List of Selected Current Geographical Serials,* University of Chicago, Department of Geography Research Paper 96 (Chicago, 1964), for further information on this matter.

5. Richard Hartshorne, *The Nature of Geography* (Lancaster, Pa.: Association of American Geographers, 1967), 247–48.

6. A great many people who have no professional concern with cartography are interested in maps, especially older decorative maps. To help satisfy an increasing demand for these, plates are removed from old atlases and mounted or framed. In addition, a new cartographic genre has arisen: the newly drawn map of antique appearance. A further development, the production of facsimile atlases and maps, is discussed in Walter W. Ristow, "Recent Facsimile Maps and Atlases," *The Quarterly Journal of the Library of Congress* (July 1967): 213–99. A recent addition to this field is the Cartart Fac T simile of Budapest, Hungary. Ronald V. Tooley founded The Map Collector's Circle in England in 1963, and a quarterly journal, *The Map Collector,* began publication in England in 1977 to cultivate this interest in old maps. See also Arthur H. Robinson, "The Potential Contribution of Cartography in Liberal Education," *Geography in Undergraduate Liberal Education* (Washington, D.C.: Association of American Geographers, 1965), 34–47, and Norman J. W. Thrower, "Cartography in University Education," *AB Bookman* 5 (1976): 5–10.

7. A particular problem for some has been the inclusion of scale in this definition. In its map cataloging procedures the Library of Congress includes as "elements" title, scale, symbols, projection, and cartographer/publishing agency, but not all of these are present in all maps. For a recent discussion of

the definition of the term *map*, see J. B. Harley and David Woodward, eds., *The History of Cartography*, vol. 1 (Chicago: University of Chicago Press, 1987), xv–xxi. To date, three volumes of this ongoing multivolume, multiauthored project have appeared: volume 1 deals with prehistoric, ancient, and early Mediterranean maps; volume 2, book 1, focuses on Islamic and south Asian maps; and volume 2, book 2 treats cartography in the traditional east and Southeast Asian societies. Also see M. J. Blakemore and J. B. Harley, *Concepts in the History of Cartography: A Review and Perspective*, Cartographica, Monograph 17 (Toronto: University of Toronto Press, 1980). Other general histories of cartography include the magisterial Leo Bagrow, *History of Cartography*, revised and enlarged by R. A. Skelton (Cambridge, Mass.: Harvard University Press, 1964), which was also published in German under the title *Meister der Kartographie* (Berlin: Safari Verlag, 1943), based on Bagrow's *Geschichte der Kartographie* of the same publisher and date; Lloyd A. Brown, *The Story of Maps* (Boston: Little Brown and Company, 1947), a librarian's view of the field; Gerald R. Crone, *Maps and Their Makers: An Introduction to the History of Cartography*, 5th ed. (Hamden, Conn.: Archon Books, 1978), a textbook written by a map curator; and John Wilford Noble, *The Mapmakers* (New York: Knopf, 1981), a popular work on the subject by a journalist. Many who have written on the history of cartography have never actually made an original, professional map, which is rather like passing a written driving test without being able to operate a car. On the other hand, practitioners are frequently so involved with a few cartographic techniques as not to be able to comprehend the larger field. Rather than citing the surveys listed here, the author has preferred to utilize original source materials or more specific studies for this book.

   8. Among the different media and techniques used for cartographic purposes by earlier and "primitive" peoples are wood, including driftwood, carved to represent relief; wooden boards, bark, skins, leather, and fabric, painted with natural dyes including blood; metal, stone, and clay marked with instruments; and clay, sand, and even snow modeled or marked with the hands. Mapmaking activities of "primitive" peoples of different geographical milieu have been discussed in the literature of the field. A collection of native cartography assembled in Russia in the early years of this century included fifty-five maps from Asia, fifteen from America, three from Africa, forty from Australia and Oceania, and two from the East Indies, as indicated in B. F. Adler, *Maps of Primitive Peoples* (St. Petersburg: Karty Piervobytnyh Narodov, 1910), and discussed in H. De Hutorowicz, "Maps of Primitive Peoples," *Bulletin of the American Geographical Society* 43 (1911): 669–79. See also Clara E. Le Gear, "Map Making by Primitive Peoples," *Special Libraries* 35, no. 3 (March 1944): 79–83; and Robert J. Flaherty, "The Belcher Islands of Hudson Bay: Their Discovery and Exploration," *Geographical Review* 5, no. 6 (June 1918): 433–43. More "advanced" peoples have also used a variety of materials in cartography: mosaic tiles, woven carpets or tapestries, painted mural decorations, globes in the form of goblets and saltcellars, and so on.

   9. Walter Blumer, "The Oldest Known Plan of an Inhabited Site Dating from the Bronze Age," *Imago Mundi* 18 (1964): 9–11. It is always dangerous to claim primacy, but this extremely early rock drawing has elicited considerable

interest. It is illustrated and discussed in George Kish, *History of Cartography* (New York: Harper and Row, 1972), 1 (number 2 of a collection of 220 map slides), and, more recently, in Catherine Delano Smith, "The Emergence of 'Maps' in European Rock Art: A Prehistoric Preoccupation with Space," *Imago Mundi* 34 (1982): 9–25.

10. Marshallese stick charts have been the subject of several articles, including Sir Henry Lyons, "The Sailing Charts of the Marshall Islanders," *Geographical Journal* 72, no. 4 (October 1928): 325–28; and William Davenport, "Marshall Island Navigation Charts," *Imago Mundi* 15 (1960): 19–26.

11. Miguel León-Portilla, "The Treasures of Montezuma," in *Maps and Map Makers,* a special issue of *The Unesco Courier* (June 1991) devoted to many aspects of the subject. This journal is published in thirty-five languages by the United Nations, an organization very much concerned with cartography. See also J. Brian Harley, "Rereading the Maps of the Columbian Encounter," *The Americas before and after 1492: Current Geographical Research, Annals of the Association of American Geographers* 82, no. 3 (September 1992): 522–42; and Louis de Vorsey Jr., "Worlds Apart: Native American World Views in the Age of Discovery," *Meridian,* Map and Geography Round Table of the American Library Association 9 (1993): 5–26, one of several places where the map referred to has been reproduced.

12. James Cooper Clark, ed. and trans., *Codex Mendoza* (London: Waterlow and Sons, Ltd., 1938), is a rare and limited facsimile edition; more sumptuous and scholarly is Frances F. Berdan and Patricia Rieff Anawalt, eds., *The Codex Mendoza,* 4 vols. (Berkeley and Los Angeles: University of California Press, 1992).

13. Ibid., 2:6.

14. The Amerindian and Inuit Maps and Mapping Programme has been established at the University of Sheffield, England, with G. Malcolm Lewis as director, and the Archive of North American Indian Maps on CD-ROM is being produced under the direction of Sona Andrews at the University of Wisconsin-Milwaukee. An exhibit titled "Cartographic Encounters: An Exhibition of Native American Maps" was mounted at the Newberry Library, Chicago, in the summer of 1993 in conjunction with the Fifteenth International Conference on the History of Cartography and was the subject of the Eleventh Kenneth Nebenzahl, Jr., Lectures on the History of Cartography, *Mapline,* a quarterly newsletter published by The Hermon Dunlap Smith Center for the History of Cartography, The Newberry Library, devoted its special issue 7 (September 1993) to "Cartographic Encounters: An Exhibition of Native American Maps from Central Mexico to the Arctic" by Mark Warhus, with illustrations and a select bibliography.

15. G. Malcolm Lewis, "The Indigenous Maps and Mapping of North American Indians," *The Map Collector* 9 (December 1979): 25–32; and other writings on this subject by the same author.

16. During the spring meeting of the California Council for Geographic Education, held at California State College at Hayward, California, on 4 May 1968, Carl O. Sauer devoted much of his banquet address to the importance of the map in geographical studies. He suggested that some maps were probably made as a pastime while others may have been used to indicate hunting and

gathering sites to show "the route to the fat oysters." In turn, a place may have received its name from the association.

17. David Turnbull, "Maps Are Territories: Science Is an Atlas," in *Nature and Human Nature* (Geelong, Victoria, Australia: Deakin University Press, 1989); Chicago: University of Chicago Press, 1993). The author would like to thank Dorothy Prescott of Melbourne for first bringing the Yolngu map to his attention.

## Chapter Two

1. G. W. Murray, "The Gold-Mine of the Turin Papyrus," *Bulletin de l'Institute d'Egypte* 24 (1941–42): 81–86.

2. The map is from Prince Youssouf (Yūsūf) Kamal, *Monumenta cartographica Africae et Aegypti*, 5 vols. in 15 fascicules (Cairo, 1926–51), where it is printed in color. This greatest of facsimile atlases, like a number of other such works, was distributed in a very limited number of sets. It is important not only for the study of Africa but also to illustrate changing cartographic forms from ancient Egypt to the period of modern exploration. See Norman J. W. Thrower, "Monumenta cartographica Africae et Aegypti," *UCLA Librarian*, supplement to 16, no. 15 (1963): 121–26; Wilhelm Bonacker, "The Egyptian Book of the Two Ways," *Imago Mundi* 7 (1965): 5–17.

3. Henry Lyons, "Two Notes on Land Measurement in Egypt," *Journal of Egyptian Archaeology* 12 (1926): 242–44.

4. Eckhard Unger, "Ancient Babylonian Maps and Plans," *Antiquity* 9 (1935): 311–22; and "From Cosmos Picture to World Maps," *Imago Mundi* 2 (1937): 1–7. See also Theophile J. Meek, "The Orientation of Babylonian Maps," *Antiquity* 10 (1936): 223–26.

5. Maps of many periods and schools, focusing on a particular area, are contained in Kamal, *Monumenta cartographica*, in which reconstructions of maps no longer extant are used in place of originals or assumed originals. The reconstructions of such maps appear in the correct chronology of the originals regardless of the date of the reconstruction. Examples of reconstructed maps focusing on Greco-Roman civilization are found in J. O. Thomson, *Everyman's Classical Atlas* (London: J. M. Dent, 1961). This readily available work contains maps of the world according to Hecataeus, Eratosthenes, Crates, and Ptolemy, some of which are derived from Sir Edward Herbert Bunbury, *A History of Ancient Geography*, 2 vols. (London: John Murray, 1883), a work of fundamental importance on the geographical and cartographical knowledge of the period.

6. In addition, purely imaginary maps have been drawn to illustrate novels and other literary works, as well as those suggested by known landscapes; an example of the first kind would be the Hobbit maps of J. R. R. Tolkien and of the second Thomas Hardy's Wessex maps. A collection of over two hundred maps of imaginary landscapes is contained in J. B. Post, *An Atlas of Fantasy*, revised ed. (New York: Ballantine Books, 1979).

7. A. E. M. Johnston, "The Earliest Preserved Greek Maps: A New Ionian Coin Type," *Journal of Hellenic Studies* 87 (1967): 86–94.

8. Bunbury, *History of Ancient Geography*, 1:615.

9. Walter W. Hyde, *Ancient Greek Mariners* (New York: Oxford University Press, 1947), 14n.

10. See Jacob Skop, "The Stade of the Ancient Greeks," *Surveying and Mapping*, 10, no. 1 (1950): 50–55.

11. See Leo Bagrow, "The Origin of Ptolemy's 'Geographia,'" *Geografiska Annaler* 27, no. 3–4 (1945): 318–87; and Johannes Keuning, "The History of Geographical Map Projections until 1600," *Imago Mundi* 12 (1955): 1–24.

12. Roman centuriation, or rectangular division of land, and its enduring effects are discussed in John Bradford, *Ancient Landscapes* (London: G. Bell, 1957), 145–216, and in George Kish, "Centuriatio: The Roman Rectangular Land Survey," *Surveying and Mapping* 22, no. 2 (1962): 233–44. A Roman surveying instrument found at Pompeii is discussed in Don Gelasio Caetini, "The 'Groma' or Cross Bar of the Roman Surveyor," *Engineering and Mining Journal-Press* (29 November 1924): 855; see also O. A. W. Dilke, "Illustrations from Roman Surveyors' Manuals," *Imago Mundi* 21 (1967): 9–29; *Greek and Roman Maps* (Ithaca: Cornell University Press, 1985); and other writings by this classical scholar on the subject of cartography in antiquity.

13. This map is reproduced in Roger J. P. Kain and Elizabeth Baigent, *The Cadastral Map in the Service of the State* (Chicago: University of Chicago Press, 1992), 2, and another is in P. D. A. Harvey, *The History of Topographical Maps: Symbols, Pictures, and Surveys* (London: Thames and Hudson, 1980), 127. One must not be misled by the main title of the latter book, which deals with many types of maps, charts, and views, generally prior to the eighteenth century.

*Chapter Three*

1. It has become fashionable, especially in the West, to minimize the contributions of colonial administrators and Occidental scholars to Oriental studies. However, Lord Curzon did much to save and make known the earlier culture of the subcontinent in founding the Archaeological Survey of India as viceroy. As an example, see Sir John H. Marshall, *Taxila: An Illustrated Account of Archeological Excavations Carried out at Taxila under Orders of the Government of India between the Years 1913 and 1934*, 3 vols. (Cambridge: Cambridge University Press, 1951). Modern maps of this archaeological site appear as plates 2, 8, 9, and 10 of volume 3, with smaller plans throughout, but no contemporary maps were found. In general, Pakistani and Indian scholars—a number of whom worked with and were trained by Marshall—have appreciated the importance of such work.

2. The most readily available English account of the cartography of China is in Joseph Needham and Wang Ling, "Mathematics and the Sciences of the Heavens and the Earth," *Science and Civilisation in China*, (Cambridge: Cambridge University Press, 1959), 3:497–590. See also E. Chavannes, "Les deux plus ancient specimens de la cartographie chinoise," *Bulletin de l'Ecole Française d'Extreme Orient* 3 (1903).

3. Kuei-Sheng Chang, "The Han Maps: New Light on Cartography in Classical China," *Imago Mundi* 31 (1979): 9–17.

4. F. Richard Stephenson, "The Ancient History of Halley's Comet," in

*Standing on the Shoulders of Giants: A Longer View of Newton and Halley,* ed. Norman J. W. Thrower (Berkeley and Los Angeles: University of California Press, 1990), 231–53.

5. The record of the contribution of women to cartography has been as shamefully neglected as that of other so-called "minorities." However, there is evidence that women played an important role in mapmaking in ancient China (see Needham and Ling, *Science and Civilisation,* 3:537–41), as, undoubtedly, they did in other countries.

6. Ibid., caption below plate 81, facing 3:548.

7. Lynn T. White, *Medieval Technology and Social Change* (Oxford: Oxford University Press, 1962), 132.

8. H. B. Hulbert, "An Ancient Map of the World," *Bulletin of the American Geographical Society* 36, no. 9 (1904): 600–605.

9. Norman J. W. Thrower and Young Il Kim, "Dong-Kook-Yu-Ji-Do: A Recently Discovered Manuscript of a Map of Korea," *Imago Mundi* 21 (1967): 10–20; Shannon McCune, "Maps of Korea," *The Far Easterly Quarterly* 4 (1948): 326–29. See also David J. Nemeth, "A Cross-Cultural Cosmographic Interpretation of Some Korean Geomancy Maps," in *Introducing Cultural and Social Cartography,* Cartographica, Monograph 44, ed. Robert A. Rundstrom (Toronto: University of Toronto Press, 1993), 85–97. There are a number of interesting articles in this collection of essays by workers cited previously (such as G. Malcolm Lewis) and some by others (Joseph E. Schwartzberg) who will be cited later. However, in his introduction Rundstrom seems to confuse "ethnic" and "indigenous" with "cultural" and "social." There is, of course, a long and continuous tradition of social and cultural cartography, as there is of social and cultural geography, and these studies hardly need "introducing."

10. Two of the earliest known indigenous Japanese maps are discussed and illustrated in Ryuziro Isida, *Geography of Japan* (Tokyo: Society for International Cultural Relations, 1961), 5–7. Occidental contributions to the mapping of the Orient before the period of modern surveys are considered in Hiroshi Nakamura, *East Asia in Old Maps* (Tokyo: Kasai, 1964). See also George H. Bean, *A List of Japanese Maps of the Tokugawa Era* (Jenkintown, Pa.: Tall Tree Library, 1951; supplements, 1955, 1958, 1963). The Bean Collection of early printed Japanese maps is now at the library of the University of British Columbia, while UCLA (Rudolph Collection) and the University of California, Berkeley (Matsui Collection) both possess great resources in this field. There is a considerable literature on the history of Japanese maps in both Japanese and European languages. Among the latter, one of the earliest is Graf Paul Teleki, *Atlas zur Geshichte der Kartographie der Japanischen Inseln* (Budapest, 1909), and one of the more recent is Hugh Cortazzi, *Isles of Gold: Antique Maps of Japan* (New York and Tokyo: Weatherhill, 1983), with a useful bibliography and many illustrations in color. See also Jason C. Hubbard, "The Map of Japan Engraved by Christopher Blancus, Rome, 1617," *Imago Mundi* 46 (1994): 84–99.

11. Reginald H. Phillimore, "Early East Indian Maps," *Imago Mundi* 7 (1950): 73–74; and "Three Indian Maps," *Imago Mundi* 9 (1952): 111–14. Susan Gole, *Early Maps of India* (New Delhi: Sanscriti, in association with Arnold-

Heineman, 1976); *India within the Ganges* (New Delhi: Jayaprints, 1983), esp. chap. 1, with illustrations; *A Series of Early Printed Maps of India in Facsimile* (New Delhi: Jayaprints, 1981); (edited), *Maps of Mughul India, Drawn by Colonel Jean-Baptiste Gentil for the French Government to the Court of Shuja-ud-Daula of Faizabad, in 1770* (New Delhi: Manohar, 1988); and *Indian Maps and Plans from Earliest Times to the Advent of European Surveys* (Tring, Herts., England: The Map Collector Publications, 1994). R. T. Fell, *Early Maps of South-East Asia,* 2d ed. (Oxford: Oxford University Press, 1991), contains European maps of Malaysia, Vietnam, the Philippines, and so forth. Joseph E. Schwartzberg, ed., *A Historical Atlas of South Asia* (Chicago: University of Chicago Press, 1978), a contemporary work with much historical data; and "A Nineteenth Century Burmese Map Relating to the French Colonial Expansion in Southeast Asia," *Imago Mundi* 46 (1994): 117–27.

## Chapter Four

1. Michael Avi-Yonah, *The Madaba Mosaic Map* (Jerusalem: Israel Exploration Society, 1954), and Herbert Donner and Heinz Cüppers, *Die Mosaikkarte von Madeba* (Wiesbaden: Otto Harrossowitz, 1977), with colored plates. M. Avi-Yonah, "The Madaba Mosaic Map," in *A Collection of Papers Complementary to the Course: Jerusalem through the Ages,* comp. Yehoshua Ben-Arieh and Shaul Sapir (Jerusalem: The Hebrew University of Jerusalem, 1984); Kenneth Nebenzahl, *Maps of the Holy Land: Images of "Terra Sancta" through Two Millennia* (New York: Abbeville, 1986); and Eran Laor, *Maps of the Holy Land: Cartobibliography of Printed Maps, 1475–1900* (New York: Alan R. Liss, Inc., 1986), which has a fine illustration of the Madaba mosaic as the book jacket.

2. Konrad Miller, *Die Peutingerische Tafel* (Stuttgart: Brockhaus, 1962), i–xii, 1–16, and half-scale color reproduction of the map. Burton William, *A Commentary of Antonius, His Itinerary, or Journies of the Romance Empire, So Far as It Concerneth Britain* (London: T. Roycroft, 1658).

3. Among prominent churchmen who apparently accepted the idea of a spherical earth were Adam of Bremen, Albertus Magnus, and Roger Bacon; those who opposed an inhabited Antipodes, if not a globular world, included Lactantius, St. Augustine of Hippo, and the oft-quoted Constantine of Antioch (Cosmos Indocopleustes), author of the *Christian Topography.* General works that discuss medieval cartography in the larger setting of the geography of the period include John K. Wright, *Geographical Lore at the Time of the Crusades* (New York: Dover, 1965), and George H. T. Kimble, *Geography in the Middle Ages* (London: Metheun, 1938).

4. This is Jerusalem; I have set her in the center of nations with countries round about her." Ezek. 5.5 RSV.

5. The Ebstorf map has appeared as a beautifully colored jigsaw puzzle with a diameter of approximately twenty inches (fifty centimeters) and an accompanying description by R(aleigh) A(shlin) (Peter) Skelton (1906–70) by Springbok Editions, New York, 1968. This excellent account of the iconography and physical characteristics of the Ebstorf map is not included in a bibliography of Skelton's published works in R. A. Skelton, *Maps: A Historical Survey of Their Study and Collecting* (Chicago: University of Chicago Press, 1972), 111–31. This important

map was destroyed in an Allied air raid in October 1943, but a hand-painted copy from facsimiles exists.

6. Gerald R. Crone, "New Light on the Hereford Map," *Geographical Journal* 131 (1965): 447–62; and *The World Map of Richard of Haldingham in Hereford Cathedral* (London: Royal Geographical Society, 1954). The latter contains a large monochrome reproduction of the Hereford map in nine sheets. Noël Denholm-Young, "The *Mappa Mundi* of Richard of Haldingham at Hereford," *Speculum* 32 (1957): 307–14. See also Waldo R. Tobler, "Medieval Distortions: The Projections of Ancient Maps," *Annals of the Association of American Geographers* 56 (1966): 351–60. This author fits a latitude and longitude graticule to the Hereford map, an interesting technique that has since been applied to other "projectionless" maps.

7. David Woodward, "Reality, Symbolism, Time and Space in Medieval World Maps," *Annals of the Association of American Geographers* 75, no. 4 (1985): 510–21. In this article the author concludes that we should evaluate the achievements of the Middle Ages in their own terms and also argues for integrating both geography and history into the study of cartography. This is certainly good advice and should be applied to maps of all types and periods.

8. Carl Schoy, "The Geography of the Moslems of the Middle Ages," *Geographical Review* 14 (1924): 257–69. In Arabic works, cartography is often discussed with geography *(djughrafiya)*, with no distinction made between the two studies. Konrad Miller, *Mappae Arabicae: Arabische Welt-und Ländeskarten der 9.-13. Jarhunderts,* 6 vols. (Stuttgart: Selbstverlag der Herausgebers, 1926–31); and *Weltkarte des Arabers Idrisi vom Jahr 1154* (Stuttgart: Brockhaus/Antiquarium, 1981). With Islam resurgent, there is now an increasing interest on the part of Muslim writers in their cultural heritage. Both scholarly and popular articles on Muslim cartography have appeared recently including Sabhi Abdel Hakim, "Atlases, Ways and Provinces," *Unesco Courier* 4 (June 1991), 20–23.

9. Anthony John Turner, "Astrolabes and Astrolabe Related Instruments," in *Early Scientific Instruments: Europe, 1400–1800* (London: Sotheby's, 1989), and Majorie Webster, Roderick Webster, and David Pingree, *Index of Western Scientific Instrument Makers to 1850* (forthcoming). Volume 1 of this compilation will deal with astrolabes and astrolabe quadrants in the Adler Planetarium Collection, Chicago. Roderick Webster, *The Astrolabe: Some Notes on its History, Construction and Use* (Lake Bluff, Ill.: Privately printed, 1984), includes materials and instructions for assembling a working model of an astrolabe.

10. Silvanus P. Thompson, "The Rose of the Winds: The Origin and Development of the Compass-Card," *British Academy Proceedings, 1913–1914* (London, 1919), 179–209; Norman J. W. Thrower, "The Art and Science of Navigation in Relation to Geographical Exploration," in *The Pacific Basin,* ed. Herman Friis (New York: American Geographical Society, 1967), 18–39, 339–43; E. G. R. Taylor and M. W. Richey, *The Geometrical Seaman: A Book of Early Nautical Instruments* (London: Hollis and Carter for the Institute of Navigation, 1962), 19–21.

11. Nils Adolf Erik Nordenskiöld, *Periplus: An Essay on the Early History of Charts and Sailing Directions* (Stockholm: P. A. Norstedt & Soner, 1897). Although some of Nordenskiöld's ideas on this subject have been discredited, this work by one of the founders of the study of historical cartography should be consulted.

See also Konrad Kretschmer, *Die italienischen Portolane des Mittlealters: Ein Beitrag zur Geschichte der Kartographie und Nautik,* vol. 13 (Berlin: Veroffentlichungen des Instituts für Meereskunde und des Geographischen Instituts an der Universität Berlin, 1909), and Edward L. Stevenson, *Portolan Charts, Their Origin and Characteristics* (New York: The Knickerbocker Press, 1911). More recent studies on portolan charts include James E. Kelley, Jr., "The Oldest Portolan Chart in the New World," *Terrae Incognitae* 9 (1977): 23–48; and Jonathan T. Lanman, "The Portolan Charts," in *Glimpses of History from Old Maps* (Tring, England: The Map Collector Publications, 1989). The last phases of portolan chartmaking in England have been discussed in Tony Campbell, "The Drapers' Company and Its School of Seventeenth Century Chart Makers," in *My Head Is a Map: A Festschrift for R. V. Tooley,* ed. Helen Wallis and Sarah Tyacke (London: Francis Edwards and Carta Press, 1973), 81–106, and in Thomas R. Smith, "Manuscript and Printed Sea Charts in Seventeenth Century London: The Case of the Thames School," in *The Compleat Plattmaker: Essays on Chart, Map, and Globe Making in England in the Seventeenth and Eighteenth Centuries,* ed. Norman J. W. Thrower (Berkeley and Los Angeles: University of California Press, 1978), 45–100. In his article (81), Campbell acknowledges the primacy of Professor Smith of the University of Kansas in discovering the important connection between late London portolan chartmakers and the Drapers' Company (guild), a link also reported by Smith in "Nicholas Comberford and the 'Thames School': Sea Chart Makers of Seventeenth Century London" (unpublished paper read at the Third International Conference in the History of Cartography, Brussels, 1969).

12. Heinrich Winter, "Catalan Portolan Maps and Their Place in the Total View of Cartographic Development," *Imago Mundi* 11 (1954): 1–12, is the last of a series of articles and a summary of additional writings by this author on the subject over a period of fifteen years. Tony Campbell, "Census of Pre–Sixteenth Century Portolan Charts," *Imago Mundi* 38 (1986): 67–94; James E. Kelley, Jr., "Non-Mediterranean Influences That Shaped the Atlantic in the Early Portolan Charts," *Imago Mundi* 31 (1979): 18–35, esp. 9 n. 4.

13. Norman J. W. Thrower, "Doctors and Maps," in *The Map Collector* 71 (summer 1995): 10–14 with illustrations.

14. Walter Horn and Ernest Born, *The Plan of St. Gall: A Study of the Architecture, Economy, and Life in a Paradigmatic Carolingian Monastery,* 3 vols. (Berkeley and Los Angeles: University of California Press, 1979), a monumental work by an art historian and a working architect.

*Chapter Five*

1. The first edition of *Maps and Man* was criticized for giving too much attention to maps of other areas and cultures—indigenous, Oriental, Islamic, colonial, and American—at the expense of European cartography. Accordingly, more discussion and also more map examples from Europe have been added, particularly to this section, to ameliorate this earlier, non-Eurocentric bias: for example, those of Juan de la Cosa and Waghenaer.

2. Thomas Goldstein, "Geography in Fifteenth-Century Florence," in *Mer-*

*chants and Scholars,* ed. John Parker (Minneapolis: University of Minnesota Press, 1965), 11–32; and Bagrow, "Ptolemy's 'Geographia.'"

3. Skelton, *Maps,* 12, who quotes William M. Ivins, Jr., on this point. *The Penrose Annual, 1964,* ed. Herbert Spencer, devotes several sections to map reproduction; for consideration of the historical aspects of this activity, see R. A. Skelton's section, "The Early Map Printer and His Problems": 171–86, and also David Woodward, ed., *Five Centuries of Map Printing* (Chicago: University of Chicago Press, 1975). This excellent work contains six essays, each by a specialist, on the principal European methods of map reproduction from early times to the present. Of course, prints are not totally alike because of wear on the plate, lines broken by use, over- or underinking, and so on, but the statement is generally true.

4. Tony Campbell, *The Earliest Printed Maps, 1472–1500* (Berkeley and Los Angeles: University of California Press, 1978). This scholarly and well-illustrated book by the map librarian of the British Library discusses and catalogs all known copies of maps printed in Europe before 1501. Also Rodney W. Shirley, *The Mapping of the World: Early Printed World Maps, 1472–1700* (London: Routledge and Paul, 1958), an invaluable compilation, especially for the map librarian and collector.

5. Actually a method of color engraving called *a la poupeé* was developed in the Netherlands by Johan(nes) Teyler in the seventeenth century but was not widely used for maps, being difficult to execute and labor-intensive. At the same time, fellows of the Royal Society were amazed at the quality of color prints from the Orient that reached London and questioned whether they were, indeed, prints or hand-colored examples.

6. Charles R. Beazley and Edgar Prestage, eds., "G. Eannes de Azurara: The Chronicle of the Discovery and Conquest of Guinea," *Hakluyt Society Publications,* series 1, vols. 95 and 100 (London: Hakluyt Society, 1896–99); Richard H. Major, *The Life of Prince Henry of Portugal, Surnamed the Navigator* (London: Asher, 1868), by the writer who popularized "Navigator" as an appellation for the "half-English Prince" who initiated Portuguese overseas voyaging; and Norman J. W. Thrower, "Prince Henry the Navigator," *Navigation* 7, no. 2–3 (1960); 117–26. See also the recent writings of Luis de Albuquerque and Alfredo Pineiro Marques on the subject of Portuguese discoveries and cartography. The primacy of Prince Pedro, rather than Prince Henry, in initiating the Portuguese overseas discoveries is now being espoused by Marques in a privately circulated newsletter, and in a book, *A Maldição da Memória Do Infante Dom Pedro* (Figueiro da Foz, 1995).

7. Goldstein, "Geography in Florence," 17–18. One of these maps is believed by some to be a copy of Marino Sanudo's *Mappa Mundi,* dated ca. 1320, on which the Azores are shown.

8. In recent years, there has been strong reaction against the claims of the European discovery of the world. A short time ago, some Zambians indicated that their ancestors "discovered" David Livingstone in the 1850s. Similarly, several Ethiopians solved the "Rhine Problem" and handed out trinkets to the natives during their ascent of that river. An early attempt to put geographical dis-

coveries in proper perspective is Friis, *The Pacific Basin* in which the exploring activities of the Pacific Islanders, Chinese, Japanese, and other non-Europeans are considered, along with those of the Portuguese, Spanish, Dutch, English, and so forth. Increasingly, indigenous peoples are adding to the information of the encounter of their ancestors with Europeans, and recently Western scholars have followed this lead, so there is now a growing body of work on the contributions of the "other" in discovery.

9. In addition to a large number of general works on exploration, the publications of the Hakluyt Society have focused on "original narratives of important voyages, travels, expeditions and other geographical records." Since its founding in 1846, the society has published approximately three hundred titles, many of which contain maps. An organization with similar objectives, the Society for the History of Discoveries, was founded in the United States in 1960; the first volume of its journal, *Terrae Incognitae*, appeared in 1969. For a popular and graphic account of the progress of European discoveries, see Leonard Outhwaite, *Unrolling the Map* (New York: John Day, 1939). Nations, especially small countries such as Portugal, are inordinately concerned with their contributions to exploration, as are others (such as New Zealand) with the particulars of their discovery. See Norman J. W. Thrower, "Cartography," in *The Discoverers: An Encyclopedia of Explorers and Exploring*, ed. Helen Delpar (New York: McGraw-Hill, 1980), 103–10.

10. R. A. Skelton, Thomas E. Marston, and George D. Painter, *The Vinland Map and the Tartar Relation* (New Haven: Yale University Press, 1965), and Wilcomb E. Washburn, ed., *The Vinland Map Conference Proceedings* (Chicago: University of Chicago Press for the Newberry Library, 1971).

11. James Enterline, *Viking America: The Norse Crossings and Their Legacy* (Garden City, N.Y.: Doubleday, 1972), and other writings by this author.

12. R. A. Skelton, *Explorers' Maps* (London: Routledge and Kegan Paul, 1958), and "Map Compilation, Production, and Research in Relation to Geographical Exploration," in Friis, *The Pacific Basin*, 40–56, 344–45. See also Armando Cortesão and Avelino Teixeira da Mota, *Portugaliae monumenta cartographica* (Lisbon: Comissão Executive das Comoracões do V Centenario da Morte do Infante D. Henrique, 1960). Among explorers' manuscript maps that have survived from the Renaissance are those of the Strait of Magellan (Antonio Pigafetta, 1524); Northern Europe (William Borough, 1570); northern New England and Nova Scotia (Samuel de Champlain, 1607); and Hudson Strait, Baffin Land, and Southampton Island (William Baffin, 1615).

13. E. G. Ravenstein, *Martin Behaim: His Life and His Globe* (London: George Philip & Son, 1908).

14. Toscanelli's map has been reconstructed and reproduced, most recently in Valerie I. J. Flint, *The Imaginative Landscape of Christopher Columbus* (Princeton: Princeton University Press, 1992), plate 40. Collections of reproductions with maps of Renaissance cartographers include Emerson D. Fite and Archibald Freeman, *A Book of Old Maps, Delineating American History from the Earliest Days Down to the Close of the Revolutionary War* (Cambridge, Mass.: Harvard University Press, 1926), including seventy-five maps reproduced in black and white with descriptions; and Kenneth Nebenzahl, *Atlas of Columbus and the Great Discoveries*

(Chicago: Rand McNally, 1990), containing splendid colored reproductions of the most important world maps of the early European overseas explorations along with explanatory text. Pizzigano's map, as well as many others discussed in this and the preceding chapter, are reproduced in Nebenzahl's *Atlas.* See also Paolo Taviani, *Christopher Columbus: The Grand Design* (London: Orbis, 1985), one of many studies by this author and others on Columbus, the quincentenary of whose first Atlantic voyage (1492–93) in 1992 generated a voluminous and sometimes vitriolic literature. Among items especially important in relation to cartography is the exhibition catalog J. B. Harley, *Maps and the Columbian Encounter* (Milwaukee: The Golda Meir Library of the University of Wisconsin, Milwaukee, 1990).

15. An important recent contribution to the literature of cartography is John P. Snyder, *Flattening the Earth: Two Thousand Years of Map Projections* (Chicago: University of Chicago Press, 1993), which, unlike most other studies on projections noted later in this book, gives a good deal of attention to early phases of this subject. See also Keuning, "Geographical Map Projections," and Norman J. W. Thrower, "Projections of Maps of Fifteenth and Sixteenth Century European Discoveries," in *Mundialización de la ciencia y culture nacional* (Madrid: Doce Calles, 1993), 81–87. See Ricardo Cerezo Martinez, "Apportión al estudio de la carta de Juan de la Cosa," in *Géographie du monde au moyen age et à la renaissance,* ed. Monique Pelletier (Paris: Editions C.T.H.S., 1989), with other articles of relevance to this chapter; and Norman J. W. Thrower, "New Geographical Horizons: Maps," in *First Images of America: The Impact of the New World on the Old,* ed. Fredi Chiappelli (Berkeley and Los Angeles: University of California Press, 1976), 659–74.

16. Bradford Swan, "The Ruysch Map of the World (1507–1508)," *Papers of the Bibliographical Society of America,* 45 (New York, 1957): 219–36. For a more recent evaluation, see Donald L. McGuirk, Jr., "Ruysch World Map: Census and Commentary," *Imago Mundi* 41 (1989): 133–41; and earlier writings by this author on the Ruysch map.

17. Edward Heawood, "The Waldseemüller Facsimiles," *Geographical Journal* 23 (1904): 760–70; Charles G. Heberman, ed., *The Cosmographiae Introduction of Martin Waldseemüller in Facsimile* (New York: The United States Catholic Historical Society, 1907).

18. George Kish, "The Cosmographic Heart: Cordiform Maps of the Sixteenth Century," *Imago Mundi* 19 (1965): 13–21, is an interesting article on this type of projection, of which a number of forms were devised by Renaissance cartographers. Recently the United States Post Office issued a "love" stamp on a cordiform projection. Many people have become interested in maps through postage stamps; cartography is one of the most popular themes in philately.

19. Piri Reis, *Kitab-I Bahriye,* 4 vols. (Istanbul: The Historical Research Foundation, Istanbul Research Center, 1988), with Piri's Atlantic chart beautifully reproduced in color as the end papers, and Gregory C. McIntosh, "Christopher Columbus and the Piri Reis Map of 1513," *The American Neptune* 53, no. 4 (fall 1993): 280–94. In this article McIntosh argues that the Piri Reis's map may be closer to the one Columbus made of his Caribbean discoveries than any other

extant chart, even that of Columbus's brother Bartholomeo with Alessandro Zorzi (1503). See also Needham and Ling, *Science and Civilisation,* 3:583–90, for influences on China.

20. Lawrence C. Wroth, *The Voyages of Giovanni da Verrazzano, 1524–1528* (New Haven: Yale University Press, 1970); and Norman J. W. Thrower, "New Light on the 1524 Voyage of Verrazzano," *Terrae Incognitae,* 11 (1979): 59–65. Verrazzano's brother Hieronymus was a cartographer who used information supplied by Giovanni in his maps.

21. Samuel Y. Edgerton, Jr., "From Mental Matrix to *Mappamundi* to Christian Empire: The Heritage of Ptolemaic Cartography in the Renaissance," in *Art and Cartography: Six Historical Essays,* ed. David Woodward (Chicago: University of Chicago Press, 1987), 10–50. See also Thrower, "New Geographical Horizons," and "When Cartography Became a Science," *Unesco Courier* (June 1991): 25–28.

22. Holbein's world map is reproduced in Ernst and Johanna Lehner, *How They Saw the New World,* ed. Gerald L. Alexander (New York: Tudor Publishing Co., 1966), 48–49. This is another compilation of many maps and pictures (in black and white) with the Vinland map in sepia as endpapers. See also *The Discovery of the World: Maps of the Earth and the Cosmos* (Chicago: University of Chicago Press, 1985), with contributions by Mrs. David Stewart, Yves Berger, Helen Wallis, and Monique Pelletier. A more recent contribution to this genre is Peter Whitfield, *The Image of the World: Twenty Centuries of World Maps* (London: The British Library, 1994), which includes many plates, mostly in color, with commentary on each of them, as well as many calendars, some of which have been expanded into books.

23. F. Van Ortroy, "Bibliographie sommaire de l'oeuvre Mercatorienne," *Revue des bibliothèques* 24 (1914): 113–48; and A. S. Osley, *Mercator: A Monograph on the Lettering on Maps, etc., in the Sixteenth-Century Netherlands* (New York: Watson Guptill Publications, 1969).

24. This topic will be dealt with later, but it is sufficient here to say that although a great deal has been written on longitude, not much has appeared on cartographic aspects of the subject. In November 1993 a conference on longitude was held at Harvard University to which the author contributed a paper, "Longitude in the Context of Cartography," in a book, *The Quest for Longitude* (forthcoming); see also Norman J. W. Thrower, "The Discovery of the Longitude," *Navigation* 5, no. 8 (1957–58): 375–81.

25. R. A. Skelton, "Hakluyt's Maps," and Helen Wallis, "Appendix: Edward Wright and the 1599 World Map," in *The Hakluyt Handbook,* ed. D. B. Quinn (London: Hakluyt Society, 1974), 1:48–73, with a reproduction of this map as an endpaper tip-in. See also Helen Wallis, "The Cartography of Drake's Voyage," in *Sir Francis Drake and the Famous Voyage, 1577–1580,* ed. Norman J. W. Thrower (Berkeley and Los Angeles: University of California Press, 1984), with reproductions of the Drake double hemispheric silver medal map by Michael Mercator, the Drake–Mellon map (in color), and other cartographic records of the first circumnavigation of the earth from which the original commander returned home.

26. Ronald V. Tooley, "California as an Island," *Map Collectors' Circle* 8 (1964), and a more recent discussion of this topic with extensive bibliography, Dora Beale Polk, *The Island of California: A History of a Myth* (Spokane, Wash.: Arthur H. Clark Company, 1991).

27. Gerardus Mercator, *Atlas sive cosmographicae meditationes de fabrica mundi et fabricati figura* (Atlas, or cosmographical meditations upon the creation of the universe and the universe as created), 3 vols. (Dusseldorf: A. Brusius, 1595). Johannes Keuning, "The History of an Atlas: Mercator–Hondius," *Imago Mundi* 4 (1947), one of a series of three articles by this scholar in this journal dealing with Mercator and his successors. Mercator–Hondius–Janssonius, *Atlas, or Geographic Description of the World*, reprinted with an introduction by R. A. Skelton, 2 vols. (Amsterdam: Theatrum Orbis Terrarum, 1968); Cornelis Koeman, *Collections of Maps and Atlases in the Netherlands* (Leiden: E. J. Brill, 1961); *A Leaf from Mercator–Hondius Atlas Edition of 1619*, with an essay in five parts by Norman J. W. Thrower (Fullerton, Calif.: Stone and Lorson, Publishers, 1985), 1–25.

28. Collectors and librarians have prized the sixteenth- and seventeenth-century productions of the cartographic houses of the Low Countries; these are often treasured as much for their aesthetic appeal as for their geographical content. See Arthur L. Humphreys, *Old Decorative Maps and Charts* (London: Halton & Truscott Smith, 1926); this was revised with a new text by R. A. Skelton as *Decorative Printed Maps of the Fifteenth to Eighteenth Centuries* (London: Staples Press, 1952), with eighty-four reproductions of maps, some in color. See Robert W. Karrow, Jr., *Mapmakers of the Sixteenth Century and Their Maps: Biobibliographies of the Cartographers of Abraham Ortelius, 1570* (Chicago: Speculum Orbis Press for the Newberry Library, 1993). For one of many substantive topics that have been investigated using early maps, see Wilma George, *Animals and Maps* (Berkeley and Los Angeles: University of California Press, 1969).

29. Günther Schilder, *Monumenta cartographica Neerlandïca*, 2 vols. and 2 portfolios (Alphen, Netherlands: Canaletto, 1986–88); and "Willem Janszoon Blaeu's Map of Europe (1606), A Recent Discovery in England," *Imago Mundi* 28 (1976): 9–20. Schilder, an Austrian now at the University of Utrecht, is the most recent of a long line of scholars interested in Low Country cartography in the Renaissance and later, including Frederik C. Wieder (who was associated with Prince Youssouf [Yusūf] Kamal in his monumental cartographic publishing ventures) and Cornelis Koeman, whose *Geschiedenis van de Kartografie van Nederland* (Alphen, Netherlands: Canaletto, 1983) should be consulted for details of the Dutch map trade and who also produced *Atlantes Neerlandici*, 5 vols. (Amsterdam: Theatrum Orbis Terrarum, 1967–71).

30. D. Gernez, "The Works of Lucas Janszoon Wagenaer [*sic*]," *The Mariner's Mirror* 23 (1937): 332–50; and Lucas Janszoon Waghenaer, *Spieghel der Zeevaerdt, Leyden, 1584–1585* (Amsterdam: Theatrum Orbis Terrarum, 1964). See also Derek Howse and Michael Sanderson, *The Sea Chart* (Newton Abbot, England: David and Charles, 1973), 40–43, 49.

31. Koeman, *Geschiedenis*, 124–26, and G. Braun and F. Hogenberg, *Civitates orbis terrarum 1572–1618* (Amsterdam: Theatrum Orbis Terrarum Ltd., 1965).

32. Tony Campbell, "The Woodcut Map Considered as a Physical Object: A New Look at Erhard Eztlaub's *Rom Weg* Map of c. 1500," *Imago Mundi* 30 (1978): 79–91.

33. Eduard Imhof, *Die Altesten Schweizerkarten* (Zurich and Leipzig: Orell Füssli Verlag, 1939).

34. Helen Wallis, ed., *The Maps and Text of the Boke of Idrography Presented by Jean Rotz to Henry VIII* (Oxford: Roxburghe Club, 1981). *Brouscon's Tidal Almanac, 1546,* with an explanation by Derek Howse and a foreword by Sir Alec Rose (Cambridge: Nottingham Court Press in association with Magdalene College, Cambridge, 1980). This is a facsimile edition of this French navigational manual with maps, once owned by Sir Francis Drake. It was later acquired by the diarist Samuel Pepys, secretary of the Royal Navy, who arranged his books according to size. *Brouscon's Almanac,* as the smallest book in Pepys's library (now at Magdalene College, Cambridge) is labeled "No. 1."

35. Sarah Tyacke and John Huddy, *Christopher Saxton and Tudor Map-Making* (London: The British Library, 1980), and William Ravenhill, *Christopher Saxton's Surveying: An Enigma,* in *English Map-Making, 1500–1560,* ed. Sarah Tyacke (London: British Library, 1983), 112–19.

*Chapter Six*

1. The five-hundredth anniversary of the birth of Copernicus was celebrated in Poland in 1973 when, among other groups, the International Society for the History of Cartography met in Warsaw. See Walter M. Brod, "Sebastian von Rotenhan, the Founder of Franconian Cartography, and a Contemporary of Nicholas Copernicus," *Imago Mundi* 27 (1975): 9–12, one of the papers read at the Warsaw conference. Both Rotenhan and Copernicus devoted their lives to the church and to science.

2. A. R. Hall, *The Scientific Revolution 1500–1800* (London: Longmans, Green and Co., 1954).

3. Galileo Galilei, *Sidereus nuncius* (Venetiis: Apud Thoman Baglionum, 1610); Judith A. Zink (Tyner), "Lunar Cartography: 1610–1962" (master's thesis, University of California, Los Angeles, 1963); and "Early Lunar Cartography," *Surveying and Mapping* 29, no. 4 (1969): 583–96.

4. William Petty, *Hiberniae delineatio* (Amsterdam, 1685; reprint, Shannon: Irish University Press, 1969). The original of this atlas was in common use until Ireland was remapped by the Ordnance Survey in the nineteenth century.

5. Helen Wallis and Arthur H. Robinson, eds., *Cartographical Innovations: An International Handbook of Mapping Terms to 1900* (Tring, England: Map Collector Publications Ltd., in association with the International Cartographic Association, 1987), an excellent compendium of map types and techniques before the advent of aerial surveys and photogrammetry. See also David A. Woodward, "English Cartography, 1650–1750: A Summary," in Thrower, *Compleat Plattmaker,* 159–93.

6. E. G. R. Taylor, "The English Atlas of Moses Pitt, 1680–83," *The Geographical Journal* 95, no. 4 (1940): 292–99. See also Norman J. W. Thrower, "The En-

glish Atlas of Moses Pitt," *UCLA Librarian* 20 (1967), and Woodward, "English Cartography," 186–87.

7. Norman J. W. Thrower, "Edmond Halley and Thematic Geo-Cartography," in *The Terraqueous Globe* (Los Angeles: William Andrews Clark Memorial Library, University of California, Los Angeles, 1969), 3–43, and "Edmond Halley as a Thematic Geo-Cartographer," *Annals of the Association of American Geographers* 59, no. 4 (1969): 652–76. See also Norman J. W. Thrower, *The Three Voyages of Edmond Halley in the "Paramore" 1698–1701*, series 2, vols. 156 and 157 (London: Hakluyt Society, 1981), the latter volume being a folio of Halley's maps arising from these voyages, and "The Royal Patrons of Edmond Halley, with Special Reference to His Maps," in Thrower, *Shoulders of Giants*, 203–19. Mr. Andrew McNally III used the Clark Library copy of Halley's Atlantic chart, published in full-color facsimile in the "first state" (without dedication), for his 1985 Christmas card. This map is one of the annual additions to this series of great interest to the student of the history of cartography.

8. For the geographical and cartographical ideas of Newton, see William Warntz, "Newton and the Newtonians, and the *Geographia Generalis Varenii*," *Annals of the Association of American Geographers* 79, no. 2 (June 1989): 165–91, with diagrams, including map projections used by Newton in his edition of Varenius's work (179).

9. Sydney Chapman, "Edmond Halley as Physical Geographer and the Story of His Charts," *Occasional Notes of the Royal Astronomical Society* 9 (London, 1941). Actually, it is a climatological chart. Physical scientists such as Chapman have been much more interested in Halley's work, including his maps, than have historians of cartography.

10. Halley felt that Mercator received too much credit for "inventing" the 1569 projection that bears his name and proposed the appellation "nautical" because of the use of this device by navigators. See Maximillian E. Novak and Norman J. W. Thrower, "Defoe and the Atlas Maritimus," *UCLA Librarian* 26 (1973), which discusses Halley's explication of the projections in Daniel Defoe's *Atlas maritimus and commercialis*.

11. Chapman, "Edmond Halley": 5.

12. It has been said that in the history of surgery there are two periods: before and after the introduction of antiseptics (by Lister). Similarly, it could be said that there are two periods in cartography, before and after the isoline, in the sense that this invention, with further development in the form of the contour, allowed the same precision for the measurement of the third dimension (height) as had previously been possible only in two dimensions (length and breadth). Of course, Halley cannot take full credit for this, since a Jesuit father in Milan, Christoforo Borri, had apparently made a manuscript magnetic chart in 1630, now lost, with entirely parallel isolines. Athanasius Kircher, who reported this in 1643, knew that it was incorrect and apparently worked on an isogonic map himself, which he intended to include with his *Magnes sive de arte magnetica opus tripartitum* (Rome, 1643). Halley refers to Kircher in his writings, but we can accept Halley's statement concerning his Atlantic chart: "What is

here properly New, is the *Curve-Lines.*" These were known as Halleyan lines for about a century, until the term *isogone* was applied to them. Kircher has the distinction of being the first to map surface ocean currents, but before the research of Alexander von Humboldt and others in the nineteenth century, the representation of these phenomena was inadequate.

13. Werner Horn, "Die Geschichte der Isarithmenkarten," *Petermanns Geographische Mitteilungen* 53 (1959): 225–32.

14. See D. W. Waters, "Captain Edmond Halley, F.R.S., Royal Navy, and the Practice of Navigation" in Thrower, *Shoulders of Giants,* 171–202. Commander Waters states: "Edmond Halley was the first and remains to this day the greatest scientific seaman." (200) With this statement Waters expanded on the opinion of Samuel Pepys (1633–1703), who averred that Halley knew more of the science and practice of navigation than anyone of his time. Sir Edward Bullard, F.R.S., described Halley as "a truly great man of prodigious versatility and most attractive personality" in his article "Edmond Halley (1656–1741)," *Endeavour* 15 (1956): 891–92; and Joseph Nicollet, whom we shall encounter later, remarked that Halley's name resounds through all of Europe and was like a clap of thunder for the philosophy of Descartes.

15. It has been suggested that this may have been the most widely distributed broadside map in England up to its time, so that the phenomenon, as Halley wrote, "may give no surprise to the People, who would, if unadvertised, be apt to look on it as Ominous." A rather neglected field of study has been the production of maps at a given period. With this in view with respect to England in the seventeenth and eighteenth centuries, Sarah Tyacke (presently Keeper of the Public Records in Britain) has analyzed advertisements in periodicals such as the *London Gazette.* See Sarah Tyacke, "Map-Sellers and the London Map Trade, c. 1650–1710," in Wallis and Tyacke, *My Head Is a Map,* 33–89.

16. S. J., Fockema Andreae, and B. van't Hoff, *Geschiedenis der Kartografie van Nederland* ('s-Gravenhage: Martinus Nijhoff, 1947).

17. Louis XIV is said to have remarked on seeing this map that he had lost more territory through the work of his hydrographic surveyors on the Atlantic coast than he had won in all his wars on the eastern frontiers of France.

18. Josef Konvitz, *Cartography in France, 1660–1848: Science, Engineering, and Statecraft* (Chicago: University of Chicago Press, 1987), a valuable and current analysis of various French contributions to cartography over a period of nearly two centuries. Among the thirty-six illustrations, eight in color, is Buache's map of the Channel (69). See also George Kish, "Early Thematic Mapping: The Work of Philippe Buache," *Imago Mundi* 28 (1976): 129–36; and R. A. Skelton, "Cartography," in *A History of Technology,* ed. Charles J. Singer (Oxford: Clarendon Press, 1954–58) 4:596–628; Buache's generalized map is reproduced (613) in this useful article. For a study of the contributions of one family to private cartography in eighteenth-century France, see Mary Sponberg Pedley, *Bel et Utile: The Work of the Robert de Vaugondy Family of Mapmakers* (Tring, England: Map Collector Publications, 1992).

19. Coolie Verner, "John Seller and the Chart Trade in Seventeenth Century England," in Thrower, *Compleat Plattmaker,* 127–57.

20. Derek Howse and Norman J. W. Thrower, eds., *A Buccaneer's Atlas: Basil Ringrose's South Sea Waggoner*, with special contributions by Tony A. Cimolino (Berkeley and Los Angeles: University of California Press, 1992). This is a facsimile of sea charts of an area about which Spain was anxious that others should not have any intelligence, and the atlas should have been jettisoned before capture. When it was taken to England, it secured from King Charles II the release of the pirates on charges that would have led to hanging, which illustrates the true "power of maps"!

21. Thomas R. Adams, *Mount and Page: Publishers of Eighteenth-Century Maritime Books*, in *A Potencie of Life: Books in Society*, ed. Nicolas Barker (London: British Library, 1993). This is one of a series of essays given at the William Andrews Clark Memorial Library, UCLA, when Barker was Clark Professor.

22. Andrew David, ed., *The Charts and Coastal Views of Captain Cook's Voyages*, vol. 1, *The Voyage of the "Endeavour," 1768–1771*, and vol. 2, *The Voyage of the "Resolution" and "Adventure" 1772–1775*, Hakluyt Society Extra Series, nos. 43 and 44 (London: 1988, 1992), contain the complete charting and topographical record of Cook's first and second Pacific voyages; volume 3, a similar record of Cook's final voyage, is forthcoming. See also Gary L. Fitzpatrick, *'Palapala'aina': The Early Mapping of Hawaii* (Honolulu: Editions Limited, 1986), esp. 12–24, with color reproductions of the maps of the islands resulting from Cook's third Pacific voyage.

23. In a highly revisionist article, "The Dimensions of the Solar System," Albert Van Helden states that "showing astronomers how Venus transits could solve the age-old problem of solar parallax [not the prediction of the periodic return of the comet that bears his name] was Halley's most fundamental contribution to astronomy." In Thrower, *Shoulders of Giants*, 143–56.

24. A great deal has been written about the astronomical and horological aspects of finding and keeping longitude at sea, but much less has appeared on the representation of meridians on maps. A book that summarizes much information on longitude, has an excellent bibliography, and adds new data on the subject is Derek Howse, *Greenwich Time and the Discovery of the Longitude* (Oxford: Oxford University Press, 1980).

25. For recent evaluations of the work of Vancouver, see Robin Fisher and Hugh Johnston, eds., *From Maps to Metaphors: The Pacific World of George Vancouver* (Vancouver: University of British Columbia Press, 1993), a series of essays arising from a conference held to mark the bicentenary of Vancouver's arrival in the Pacific Northwest to continue charting begun on Cook's third voyage. This volume contains maps and charts but none by Vancouver, unlike a highly illustrated work, Robin Fisher, *Vancouver's Voyage: Charting the Northwest Coast, 1791–1795* (Seattle: University of Washington Press, 1992), which includes several of Vancouver's maps and views. A popular book on the first circumnavigation of Australia is K. A. Austin, *The Voyage of the "Investigator," 1801–1803: Commander Matthew Flinders, R.N.* (Adelaide: Rigby, Ltd., 1964), with details from Flinders's charts.

26. Adrian H. W. Robinson, *Marine Cartography in Britain* (Leicester: Leicester University Press, 1962), and Sir John Edgell, *Sea Surveys* (London: H.M. Sta-

tionery Office, 1965), for general accounts of Britain's role in hydrographic charting.

27. Leo Bagrow, *A History of the Cartography of Russia*, ed. Henry W. Castner, 2 vols., (Wolfe Island, Ont.: The Walker Press, 1975). This is a translation of a monumental work by Bagrow, one of the modern founders of the study of early maps, copiously illustrated with black-and-white reproductions of maps of Russia by both foreign and indigenous Russian cartographers.

28. François de Dainville, "De la profondeur à l'altitude," *International Yearbook of Cartography* 2 (1962): 151–62, translated into English and published as "From the Depths to the Heights," *Surveying and Mapping* 30 (1970): 389–403; also Arthur H. Robinson, "The Geneology of the Isopleth," *Cartographic Journal* 8 (1971): 49–53.

29. Sir Herbert George Fordham, *Some Notable Surveyors and Map-Makers of the Sixteenth, Seventeenth, and Eighteenth Centuries and Their Work* (Cambridge: Cambridge University Press, 1929), esp. chap. 3. See also Lloyd A. Brown, *Jean-Dominique Cassini and His World Map of 1696* (Ann Arbor: University of Michigan Press, 1941), and Konvitz, *Cartography in France*, esp. chap. 1.

30. Elia M. J. Campbell, "An English Philosophico-Chorological Chart," *Imago Mundi* 6 (1950): 79–84.

31. M. Foncin, "Dupin-Triel [*sic*] and the First Use of Contours," *The Geographical Journal* 127 (1961): 553–54.

32. Tim Owen and Elaine Pilbeam, *Ordnance Survey: Map Makers to Britain Since 1791* (London: H.M. Stationery Office, for the Ordnance Survey, 1992), and R. A. Skelton, "The Origins of the Ordnance Survey of Great Britain," pt. 3 of "Landmarks in British Cartography," *The Geographical Journal* 78 (1962): 406–30.

33. Matthew H. Edney, "The Patronage of Science and the Creation of Imperial Space: The British Mapping of India, 1799–1843," *Cartographica* (1993): 61–67; another view of this great enterprise is given in "Measuring India," in *The Shape of the World*, (Chicago, New York, San Francisco: Rand McNally, 1991), chap. 8. This collection arose from a public television series of the same name dealing with geographical discovery and with mapping from antiquity to the present, and it contains excellent color and black-and-white illustrations.

34. Helen Wallis, "Raleigh's World," in *Raleigh and Quinn: The Explorer and His Boswell*, ed. H. G. Jones (Chapel Hill: North Carolinian Society, Inc., and the North Carolina Collection, 1987), 11–33; and John P. H. Hulton and David B. Quinn, *The American Drawings of John White, 1577–1590* (London and Chapel Hill: British Museum and University of North Carolina Press, 1964).

35. P. E. H. Hair, "A Note on Thevet's Unpublished Maps of Overseas Islands," *Terrae Incognitae* 14 (1982): 105–16.

36. Walter W. Ristow, "Augustine Hermann's Map of Virginia and Maryland," *Library of Congress Quarterly Journal of Acquisitions* (August 1960): 221–26.

37. Coolie Verner, *The Fry & Jefferson Map of Virginia and Maryland* (Charlottesville: University Press of Virginia, 1966).

38. Seymour J. Schwartz and Ralph E. Ehrenberg, *The Mapping of America* (New York: Harry N. Abrams, Inc., 1980), 96, 159–60, an excellent survey of the

cartography of North America from earliest times to the space age. Part one by Schwartz deals with the subject from 1500 to 1800, part two by Ehrenberg, 1800 to the present. There are a number of good regional histories of cartography, such as William P. Cumming, *The Southeast in Early Maps* (Princeton: Princeton University Press, 1958), and other works on this subject by this author. See also Jeannette D. Black, "Mapping the English Colonies: The Beginnings," in Thrower, *Compleat Plattmaker,* 101–27, with a pen-and-ink map of the Cape Fear River on the Carolina coast by the English philosopher John Locke; also *The Blathwayt Atlas: A Collection of Forty-eight Manuscript and Printed Maps of the Seventeenth Century,* ed. Jeannette D. Black (Providence: Brown University Press, 1970).

39. Louis de Vorsey, "Pioneer Charting of the Gulf Stream: The Contributions of Benjamin Franklin and William Gerard De Brahm," *Imago Mundi* 28 (1976): 105–20.

40. Coolie Verner, "Mr. Jefferson Makes a Map," *Imago Mundi* 14 (1959): 96–108. Jefferson declared that this map, which was based on the Fry–Jefferson map and included in his *Notes on the State of Virginia,* was "of more value than the book in which it appeared."

41. Herman R. Friis, "A Brief Review of the Development and Status of Geographical and Cartographical Activities of the United States Government: 1776–1818," *Imago Mundi* 19 (1965): 68–80; and J. B. Harley, Barbara Bartz Petchenik, and Lawrence W. Towner, *Mapping the American Revolutionary War* (Chicago: University of Chicago Press, 1978).

42. Information on map projections, including methods of construction, is contained in the standard textbooks on cartography listed earlier and in a number of special publications. Among the latter, outstanding works are Charles H. Deetz and Oscar S. Adams, *Elements of Map Projections* (Washington, D.C.: U.S. Coast and Geodetic Survey, 1934); Irving Fisher and Osborn M. Miller, *World Maps and Globes* (New York: Essential Books, 1944); and James A. Steers, *An Introduction to the Study of Map Projections,* 13th ed. (London: University of London Press, 1962). See also Snyder, *Flattening the Earth,* esp. 55–94, for this period.

43. Johann H. Lambert, *Beyträge zum Gebrauche der Mathematick und deren Anwendung,* 5 pts. (Berlin: Reimer, 1765–72).

44. The memory of Coronelli is honored in the name of an international society: Coronelli–World League of Friends of the Globe. See Helen Wallis, "Geographie Is Better Than Divinitie: Maps, Globes and Geographie in the Days of Samuel Pepys," in Thrower, *Compleat Plattmaker,* 1–43.

45. As was not uncommon at the time, a number of women continued the businesses of their husbands after these were deceased, including the widow of Jodocus Hondius, Coletta van den Keere, who was assisted by her sons, Jodocus II and Henricus. A somewhat different case is that of Virginia Farrer (Farrar, Ferrar, and other variants), daughter of John Farrer, cartographer and deputy treasurer of Virginia in the mid–seventeenth century, who issued later versions of her father's map of Virginia under her own name. The Farrer map of Virginia (London, 1651) shows Raleigh's Roanoke colony on the Atlantic (now North Carolina) to be ten days' march away from Drake's Nova Albion (California) on

the Pacific coast! See Ronald V. Tooley, *Tooley's Dictionary of Map Makers* (Tring, England: Map Collector Publications, 1979), 203; and Helen Wallis, (with reproductions of the map) in Norman J. W. Thrower, *Sir Francis Drake and the Famous Voyage, 1577–1580* (Berkeley and Los Angeles: University of California Press, 1984), 157–58. For a branch of cartography that is virtually a female monopoly, see Judith Tyner, "Geography through the Needle's Eye: Embroidered Maps and Globes in the Eighteenth and Nineteenth Centuries," *The Map Collector* 66 (spring 1994): 2–7, with illustrations, some in color. On another note, at the time of this writing (February 1995) the U.S. space shuttle, on close approach to the Russian space station *Mir*, is being navigated by Eileen Collins.

*Chapter Seven*

1. William Petty, *Political Arithmetick* (London: R. Clavel and Hortlock, 1690); Christian Huygens, *De ratiociniis in ludo aleae* (1657); Edmond Halley, "Contemplation on the Mortality of Mankind," *Philosophical Transactions* (1693), 17:596ff.

2. Skelton, "Early Map Printer": esp. 182–84, including a color plate, and L. Dudley Stamp, "Land Use Surveys with Special Reference to Britain," in *Geography in the Twentieth Century*, ed. Griffith Taylor (London: Methuen, 1951), 372–92.

3. Kirtley F. Mather and Shirley L. Mason, *A Source Book in Geology* (New York: McGraw-Hill, 1939), esp. 201–4.

4. See Josef Szaflarski, "A Map of the Tatra Mountains Drawn by George Wahlenberg in 1813 as a Prototype of the Contour-line Map," *Geografiska Annaler* 41 (1959): 74–82; and Ingrid Kretschmer, "The First and Second School of Layered Relief Maps in the Nineteenth and Early Twentieth Centuries," *Imago Mundi* 40 (1988): 9–14, with colored frontispiece.

5. See Humboldt's own voluminous writings and, especially for this period of his life, his *Political Essay on the Kingdom of New Spain* (1811; reprint, New York: Knopf, 1972), one of a number of reprints of this classic work in English. See also Norman J. W. Thrower, "Humboldt's Mapping of New Spain (Mexico)," *The Map Collector* 53 (winter 1990): 30–35. In his *Political Essay*, Humboldt wrote: "The indication of the chains of mountains presented difficulties which can only be felt by those who have themselves been employed in the construction of geographic maps. I prefer hatchings [hachures] in orthographical projection to the method of representing in profile."

6. Arthur H. Robinson and Helen M. Wallis, "Humboldt's Map of Isothermal Lines: A Milestone in Thematic Cartography," *The Cartographic Journal* 4, no. 2 (1967): 119–23; and Arthur H. Robinson, *Early Thematic Cartography in the History of Cartography* (Chicago: University of Chicago Press, 1982), the first overview of this subject by the pioneer in such studies.

7. Norman J. W. Thrower, ed., *Man's Domain: A Thematic Atlas of the World. Mapping Man's Relationship with His Environment*, 3d ed. (New York: McGraw-Hill, 1975), 4.

8. Herman R. Friis, "Highlights of the History of the Use of Conventionalized Symbols and Signs on Large-scale Nautical Charts of the United States Gov-

ernment," *1st Congrès d'Histoire de l'Océanographie, Bulletin de l'Institute Océano-graphique,* numero special 2 (Monaco, 1968): 223–41; and "Brief Review."

9. Szaflarski, "Tatra Mountains": 75; Owen and Pilbeam, *Ordnance Survey,* 50.

10. Martha Coleman Bray, *Joseph Nicollet and His Map* (Philadelphia: The American Philosophical Society, 1980).

11. Norman J. W. Thrower, "William H. Emory and the Mapping of the American Southwest Borderlands," *Terrae Incognitae: The Journal for the History of Discoveries* 22 (1990): 41–91.

12. A. Philip Muntz, "Union Mapping in the American Civil War," *Imago Mundi* 17 (1963): 90–94, with eight contemporary Civil War maps; and George B. Davis et al., *The Official Military Atlas of the Civil War* (New York: Arno Press, Inc., 1978), with general maps, photographs (views), and drawings.

13. William D. Pattison, *Beginnings of the American Rectangular Land Survey System, 1784–1800,* University of Chicago, Department of Geography Research Paper 50 (Chicago, 1957), and Francis J. Marschner, *Boundaries and Records in the Territory of Early Settlement from Canada to Florida* (Washington, D.C.: U.S.D.A. Agricultural Research Service, 1960). Ronald E. Grim, "Maps of the Township and Range System," in *From Sea Charts to Satellite Images: Interpreting North American History through Maps* ed. David Buisseret (Chicago: University of Chicago Press, 1990), chap. 4. This treasure trove treats a dozen different aspects of the representation of the American landscape in different media over several centuries by a number of authorities.

14. Richard W. Stephenson, *Land Ownership Maps: A Checklist of Nineteenth Century United States County Maps in the Library of Congress* (Washington, D.C.: Library of Congress, 1967). Cartobibliographical studies such as this are of great value to the research scholar. Sometimes the works are topical, as in the case above, or regional, as for example in the American cartobibliographic studies of William Cumming, Carl Wheat, and Henry Wagner.

15. Clara E. Le Gear, "United States Atlases: A List of National, State, County, City and Regional Atlases in the Library of Congress" (Washington, D.C.: Library of Congress, 1950), and "United States Atlases: A Catalog of National, State, County and Regional Atlases in the Library of Congress and Cooperating Libraries" (Washington, D.C.: Library of Congress, 1953). Some four thousand different United States county atlases are listed in these two volumes. Norman J. W. Thrower, "The County Atlas of the United States," *Surveying and Mapping* 21, no. 3 (1961): 365–72; Michael Conzen, "The County Landowner-ship Map in America: Its Commercial Development and Social Transformation, 1814–1939," *Imago Mundi* 36 (1984): 9–31; John F. Rooney, Wilbur Zelinsky, and Dean R. Louder, eds., *This Remarkable Continent: An Atlas of United States and Canadian Society and Culture* (College Station, Tex.: Texas A & M University Press, 1982), esp. Terry G. Jordan, "Division of Land," 54–69.

16. Norman J. W. Thrower, "Cadastral Surveys and County Atlases of the United States," *The Cartographic Journal* 9 (1972): 43–51. In giving the author permission to quote a long passage from her late husband's work, Vernice Lock-ridge Noyes mentioned that *An Historical Atlas of Henry County,* Indiana (Chicago:

Higgins Beldon and Co., 1875) was the model referred to in *Raintree County.*

17. The late Sir H. Clifford Darby, editor of the monumental *Domesday Geography of England,* once remarked to the author of this book that the county atlas is the "American Domesday" in terms of its potential value for studies in historical geography in the United States and parts of Canada. Norman J. W. Thrower, *Original Survey and Land Subdivision: A Comparative Study of the Form and Effect of Contrasting Cadastral Surveys* (Chicago: Association of American Geographers, 1966). This is a study in historical geography and cultural cartography based largely on county atlases and plat books, ca. 1875 and ca. 1955, covering two one-hundred-square-mile areas in Ohio. See also Hildegard Binder Johnson, *Order upon the Land: The U.S. Rectangular Land Survey in the Upper Mississippi* (New York: Oxford University Press, 1976).

18. Walter W. Ristow, "United States Fire Insurance and Underwriters Maps: 1852–1968," *The Quarterly Journal of the Library of Congress* 25, no. 3 (July 1968): 194–218; and *American Maps and Mapmakers: Commercial Cartography in the Nineteenth Century* (Detroit: Wayne State University Press, 1985), a splendid overview of many aspects of private cartography and cartographers in the United States by the former chief of the Geography and Map Division of the Library of Congress.

19. James Elliot, *The City in Maps: Urban Mapping to 1900* (London: The British Library, 1987), an excellent catalog of urban maps with many illustrations, a number in color. Earlier representations of cities, especially in the United States, are available in facsimile and are listed in "Historic City Plans and Views" published annually in Ithaca, N.Y.; and John R. Hébert and Patrick E. Demsey, *Panoramic Maps of Cities in the United States and Canada,* 2d ed. (Washington, D.C.: Library of Congress, 1984), a checklist of such items with black-and-white illustrations and bibliography. Perspective is very much a product of the European Renaissance and is not characteristic of the art or cartography of other areas, such as the Orient.

20. Melville C. Branch, *Comparative Urban Design: Rare Engravings, 1830–1843* (New York: Arno Press, 1978). This work reproduces a number of SDUK engravings and provides valuable explication of them. See also Mead T. Cain, "The Maps of the Society for the Diffusion of Useful Knowledge: A Publishing History," *Imago Mundi* 46 (1994): 151–67.

21. Woodward, *Five Centuries,* esp. Walter W. Ristow, "Lithography and Maps, 1796–1850," and Cornelis Koeman, "The Application of Photography to Map Printing and the Transition to Offset Lithography."

22. Andrew M. Modelski, *Railroad Maps of North America: The First Hundred Years* (Washington, D.C.: Library of Congress, 1984). See also David Woodward, *The All-American Map: Wax Engraving and Its Influence on Cartography* (Chicago: University of Chicago Press, 1977).

23. Howse, *Greenwich Time,* esp. 116–71, with a picture of Charles F. Dowd, the American who devised the global time-zone system and maps of this for the United States and the world.

24. Matthew Fontaine Maury, *Explanations and Sailing Directions to Accompany the Wind and Current Charts,* 7th ed. (Philadelphia: E. C. and J. Biddle, 1855);

Charles Lee Lewis, *Matthew Fontaine Maury: The Pathfinder of the Seas* (Annapolis: The United States Naval Institute, 1927).

25. Henry Stommel, *Lost Islands: The Story of Islands That Have Vanished from Nautical Charts* (Vancouver: University of British Columbia Press, 1984). This book by a senior scientist at the Woods Hole Oceanographic Institute should serve as a caution to those who would put too much faith in printed matter, whether books or maps and charts.

26. Arthur H. Robinson, "The 1837 Maps of Henry Drury Harness," *The Geographical Journal* 121 (1955): 440–50. Soon after this seminal study appeared, the late R. A. Skelton, at the time superintendent of the Map Room at the British Museum, remarked to the author of this book, "That article certainly opened some eyes over here [in Britain], and we now always show Harness's atlas to groups of university students who come to the Map Room." In effect, Robinson's article had created a new field of study, that of thematic maps.

27. This technique is discussed and illustrated in John K. Wright, "A Method of Mapping Densities of Population with Cape Cod as an Example," *The Geographical Review* 26 (1936): 103–10.

28. E. W. Gilbert, "Pioneer Maps of Health and Disease in England," *The Geographical Journal* 124 (1958); and Thrower, "Doctors and Maps." A large-scale reproduction of part of Dr. Snow's map that shows the original symbolization is found in Elliot, *The City in Maps*, 80.

29. Arthur H. Robinson, "The Thematic Maps of Charles Joseph Minard," *Imago Mundi* 21 (1967): 95–108.

30. John K. Wright, "The Field of the Geographical Society," in G. Taylor, *Geography in the Twentieth Century*, 543–65. This general work includes references to the cartography of the geographical societies and lists in a footnote individual histories of a number of these societies. It also contains articles on various aspects of cartography—cultural, social, political, and so forth—as inherited from the nineteenth century and practiced in the first half of the twentieth century.

31. John N. L. Baker, *A History of Geographical Discovery and Exploration* (Boston: Houghton Mifflin, 1931). Though somewhat out of date, with too "colonial" an outlook for most contemporary scholars, this book is still a good one-volume narrative summary of geographical discovery, including the exploration of the interiors of the continents as well as coastal areas. See also Delpar, *The Discoverers*.

32. Richard A. Bartlett, *Great Surveys of the American West* (Norman, Okla.: University of Oklahoma Press, 1962), and William H. Goetzmann, *Exploration and Empire: The Explorer and Scientist in the Winning of the American West* (New York: Knopf, 1966). Robert H. Block, "The Whitney Survey of California, 1860–1874: A Study of Environmental Science and Exploration" (Ph.D. diss., University of California, Los Angeles, 1982), shows the pioneering role of Whitney in scientific surveys in the American West. Two recently published works with many illustrations of maps of the area west of the Mississippi River are Frederick C. Luebke, Francis W. Kaye, and Gary E. Moulton, eds., *Mapping the North American Plains: Essays in the History of Cartography* (Norman, Okla.: University of Oklahoma Press, 1983), and Dennis Reinhartz and Charles C. Colley, eds., *The Mapping of the American Southwest* (College Station, Tex.: Texas A & M University Press, 1987).

33. The American Civil War was the first conflict to be thoroughly covered by photography, and techniques learned in the field were soon employed for topographic illustration. Nevertheless, the topographic artist still played a vital role, as shown by the splendid drawings that illustrate the reports of surveys of the American West from the 1870s.

34. Snyder, *Flattening the Earth*, reproduces portraits of a number of those who contributed to projections, as well as illustrating and discussing their work.

35. Armin K. Lobeck, *Block Diagrams and Other Graphic Methods Used in Geology and Geography*, 2d ed. (Amherst: Emerson-Trussel, 1958); Norman J. W. Thrower, "Block Diagrams and Mediterranean Coastlands: A Study of the Block Diagram as a Technique for Illustrating the Progressive Development of Low to Moderately Sloping Coastlands of the Mediterranean Region" (B.A. thesis, University of Virginia, 1953).

36. Wright, "Geographical Society."

37. Friis, "Brief Review": 78, 80.

38. The map as an instrument of statecraft from the sixteenth to the nineteenth century was the subject of a cartographic exhibit at the Newberry Library, Chicago, titled "Monarchs, Ministers and Maps," with an illustrated catalog of the same title (Chicago: Newberry Library, 1985). The catalog was prepared by James Akerman and David Buisseret with the assistance of Arthur Holzheimer.

## Chapter Eight

1. Understandably, photography has a large literature, but a particularly valuable account in relation to mapping is contained in Robert N. Colwell, ed., *Manual of Photographic Interpretation* (Washington, D.C.: American Society of Photogrammetry, 1960), esp. 1–18. Colwell is one of the pioneers in the field of remote sensing of the environment, and his work was expanded into *Manual of Remote Sensing*, 2d ed., 2 vols. (Falls Church, Va.: American Society of Photogrammetry, 1983), with David Simonett and John Estes as special editors of the respective volumes, which include contributions by many of the important scholars in the field.

2. Beaumont Newhall, *Airborne Camera: The World from the Air and Outer Space*, (New York and Rochester, N.Y.: Hastings House Publishers in collaboration with the George Eastman House, 1969), includes a reproduction of George Catlin's painting "The Topography of Niagara" and many illustrations of landmark photographs in the history of aerial mapping and remote sensing, a term that was coined because *airphoto interpretation* (API) did not include more exotic imaging practices and output such as radar, infrared, ultraviolet, and passive microwave.

3. Federal Republic of Germany, *United Nations Technical Conference on the International Map of the World on the Millionth Scale* (Bonn: Institut für Angewandte Geödasie, 1962); Hans-Peter Kosack, "Cartographic Problems of Representing the Polar Areas on the International Map of the World on the Scale 1:1,000,000 and on Related Series," a background paper for the United Nations, *Technical Conference on the International Map of the World on the Millionth Scale* (Bonn, 1962). The United Nations also sponsors thematic mapping such as UNESCO's "Bioclimatic Map of the Mediterranean Region" by Henri Gaussen, a leader in the field

of vegetation mapping, and others. The United Nations hosts regional conferences on mapping problems in many different parts of the world and publishes reports on these, such as the "United Nations Regional Cartographic Conference for Asia and the Far East" (United Nations, Department of Economic and Social Affairs, 1977) and the "United Nations Regional Conference of the Americas" (United Nations, Department of Technical Cooperation for Development, 1989).

4. United Nations, "First Progress Report on the International Map of the World on the Millionth Scale (1954)," *World Cartography* 4 (1954).

5. Richard A. Gardiner, "A Re-Appraisal of the International Map of the World (IMW) on the Millionth Scale," *International Yearbook of Cartography* 1 (1961): 31–49.

6. International Civil Aviation Organization, *Aeronautical Information Provided by States*, 22d ed. (Quebec: International Civil Aviation Organization, 1968).

7. The Hydrographic Bureau publishes *The International Hydrographic Bulletin, The International Hydrographic Review, International Hydrographic Organization Yearbook*, and other valuable material on world charting on a regular basis. See also G. S. Ritchie, *No Day Too Long: An Hydrographer's Tale* (Durham: The Pentland Press, 1992), an autobiography by Rear Admiral Ritchie, nineteenth hydrographer of the Royal Navy and for ten years at the International Hydrographic Bureau, Monaco; and *The Admiralty Chart: British Naval Hydrography in the Nineteenth Century*, revised ed. (Durham: The Pentland Press, 1995).

8. Thomas M. Lillesand and Ralph W. Kiefer, *Remote Sensing and Image Interpretation*, 3d ed. (New York: Wiley, 1993), is an excellent one-volume treatment of many ideas touched on in this chapter, including photogrammetry and aerial and satellite image interpretation and applications. George S. Whitmore, Morris M. Thompson, and Julius L. Speert, "Modern Instruments for Surveying and Mapping," *Science* 130 (1959): 1059–66; and Morris M. Thompson, *Maps for America: Cartographic Products of the United States Geological Survey and Others*, 2d ed. (U.S. Department of the Interior, Geological Survey, 1981). This work contains a short history of the USGS and samples of the various maps of the survey, many in color, and it illustrates principal modern methods of mapmaking with pictures of instruments. For illustrations of official maps of Britain with text, see J. B. Harley, *Ordnance Survey Maps: A Descriptive Manual* (Southampton: Ordnance Survey, 1975); this author has also described reprinted one-inch Ordnance Survey maps of historical interest.

9. Norman J. W. Thrower and John R. Jensen, "The Orthophoto and Orthophotomap: Characteristics, Development and Application," *The American Cartographer* 3 (April 1976): 39–56; and C. D. Burnside, *Mapping from Aerial Photographs*, 2d ed. (New York: Wiley, 1985), one of a number of books on this subject, including the production of orthophotomaps. A symposium titled "Photo Maps and Orthophotomaps" was held in Ottawa, Canada, in 1967, the proceedings of which were published in *The Canadian Surveyor* 22, no. 1 (1968): 1–220.

10. Eduard Imhof, *Cartographic Relief Presentation*, ed. H. J. Steward (Berlin and New York: Walter de Gruyter, 1982). This is the English edition of a classic work on terrain representation, including subjective methods, by the Swiss mas-

ter of this art, who was also the founder of the International Cartographic Association (ICA).

11. United States Geological Survey, "Topographic Maps—Descriptive Folder" (Washington, D.C.: USGS, regularly revised). The USGS, like other great surveys of the world, publishes status maps of topographic coverage, aerial photo coverage, geodetic control, and so forth, on a regular basis.

12. H. Arnold Karo, *World Mapping, 1954–55* (Washington, D.C.: Industrial College of the Armed Forces, 1955). Although now out of date, this book has an excellent series of world "appraisal" maps of triangulation as well as coverage of topographic maps, nautical charts, aeronautical charts, and so forth, of the mid–twentieth century. Also, for the same period, Everett C. Olson and Agnes Whitmarsh, *Foreign Maps* (New York: Harper, 1944), which illustrates major sheet maps of the world by small samples, some in color. See also C. B. Muriel Lock, *Modern Maps and Atlases: An Outline Guide to Twentieth Century Production* (Hamden, Conn.: Archon Books, 1969). This work, unlike the Olson and Whitmarsh volume, lacks illustrations except for a frontispiece. The above works are of considerable historical interest but have been replaced for modern global, cartographic information by R. B. Parry and C. R. Perkins, *World Mapping Today* (London: Butterworths, 1987) and Rolf Böhme, comp., *Inventory of World Topographic Mapping*, 3 vols. (London and New York: International Cartographic Association, 1991). These volumes have small samples of maps in black and white, coverage diagrams, and so on. The International Cartographic Association is the principal international professional society and clearinghouse for contemporary cartographic ideas and information, with major meetings held in different parts of the world every three years.

13. Map reading, measurement, analysis, intelligence, and so forth have a considerable literature, most of which emphasizes the topographic map. Good representative examples are the *Department of the Army Field Manual 21–26* and T. W. Birch, *Maps, Topographical and Statistical*, 2d ed. (Oxford: The Clarendon Press, 1964). An interesting departure from such traditional map-reading manuals is Armin K. Lobeck, *Things Maps Don't Tell Us* (New York: Macmillan, 1956); reprint, Chicago: University of Chicago Press, 1990). David Greenhood, *Mapping* (Chicago: University of Chicago Press, 1964) deals with the technique of mapmaking from a topical point of view. See also Norman J. W. Thrower and Ronald U. Cooke, "Scales for Determining Slope from Topographic Maps," *The Professional Geographer* 20, no. 3 (1968). 181–86. One hundred 1:62, 500 quadrangles were specially selected from the USGS topographic coverage of the United States to illustrate physiographic types. See also *Rural Settlement Patterns in the United States as Illustrated on One Hundred Topographic Quadrangle Maps*, pub. 380 (National Academy of Sciences, National Research Council, 1956), as well as Tyner, *World of Maps;* Muehrcke and Muehrcke, *Map Use;* and Monmonier and Schnell, *Map Appreciation*. An outstanding work showing the relationship of photo to map is *Nouvel atlas des formes du relief* (Paris: Nathan, 1985), a revision of M. Cholley, ed., *Un atlas des formes du relief* (1956).

14. Richard G. Ray, *Photogeologic Procedures in Geologic Interpretation and Mapping: Procedures and Studies in Photogeology*, Geological Survey Bulletin 1043-A

(Washington, D.C.: United States Government Printing Office, 1956), and a more recent manual, John Barnes, *Basic Geological Mapping*, 2d ed. (New York and Toronto: Halsted Press, 1991). For more general coverage, Joseph McCall and Brian Marder, eds., *Earth Science Mapping for Planning, Development and Conservation* (London: Graham and Trotman, 1989).

15. L. Dudley Stamp and E. C. Willatts, *The Land Utilisation Survey of Britain*, 2d ed. (London: London School of Economics, 1935). A verbal description of the survey was published in ninety two parts between 1937 and 1941. A work on a related topic is A. W. Küchler and I. S. Zonneveld, eds., *Vegetation Mapping* (Dordrecht, Boston, London: Kluwer Academic Publishers, 1988). The United States Geological Survey has published selected land-use and land-cover maps recognizing nine principal and thirty-eight secondary categories.

16. Norman J. W. Thrower, *Satellite Photography as a Geographic Tool for Land Use Mapping in the Southwestern United States* (Washington, D.C.: United States Geological Survey, 1970); Norman J. W. Thrower assisted by Leslie W. Senger and Robert H. Mullens, cartography by Carolyn Crawford and Keith Walton, "Land Use in the Southwestern United States from Gemini and Apollo Imagery," map supplement no. 12, *Annals of the Association of American Geographers* 60, no. 1 (March 1970). This project, funded by the USGS, resulted in one of the earliest such maps, the number of which has grown considerably in recent years. The images from which the map was made overlapped the international boundary uniformly, unlike map series of Mexico and the United States that are different in date, scale, symbolization, and even system of measurement used. An attempt to overcome this problem is Norris Hundley and Norman J. W. Thrower, eds., *United States/Mexico Borderlands Atlas* (forthcoming). See J. Denègre, ed., *Thematic Mapping from Satellite Imagery: A Guidebook* (Tarrytown, N.Y.: Elsevier for the International Cartographic Association, 1994), a work also available in French. For uniform cartographic coverage of two widely separated but geographically similar areas of the world, see Norman J. W. Thrower and David E. Bradbury, eds., *Chile–California Mediterranean Scrub Atlas: A Comparative Atlas* (Stroudsburg, Pa.: Dowden, Hutchinson and Ross, Inc., for the International Biological Program [IBP], 1977).

17. There is an abundant literature on remote sensing, both professional and popular. The latter includes both imagery atlases and the "from above" coffee-table books; the former, Colwell, *Remote Sensing*, and John R. Jensen, *Introductory Digital Image Processing* (Englewood Cliffs, N.J.: Prentice-Hall, 1986). A valuable textbook on the subject is Floyd B. Sabins, Jr., *Remote Sensing: Principles and Interpretation*, 2d ed. (New York: W. H. Freeman, 1987), while Nicholas M. Short, *The Landsat Tutorial Workbook: Basics of Remote Sensing* (Washington, D.C.: National Aeronautics and Space Administration, 1982), is a useful aid in courses, with a good bibliography.

18. Ronald J. Wasowski, "Some Ethical Aspects of International Satellite Remote Sensing," *Photogrammetric Engineering and Remote Sensing* 57 (1991): 41–48. This journal is a major vehicle for information in these fields.

19. Denis Wood, with John Fels, *The Power of Maps* (New York: The Guilford Press, 1992), esp. 48–61. Studies have been made of migration and more local

movements of birds, animals, and insects. These suggest complicated spatial "instincts" that remain as enigmatic to humans as a map or remote-sensing image might be to these creatures.

20. Mark Monmonier, "Telegraphy, Iconography, and the Weather Map: Cartographic Reports by the U.S. Weather Bureau, 1870–1937," *Imago Mundi* 40 (1988): 15–31. In this section of the book, the author was greatly assisted by Professor James K. Murakami of the UCLA Department of Atmospheric Sciences.

21. Oliver E. Baker, *Atlas of American Agriculture* (Washington, D.C.: United States Department of Agriculture, 1936); also Marschner, *Boundaries and Records.*

22. The work of census organizations is eminently geographical and cartographic, as illustrated by *Census '90 Basics* (Washington, D.C.: Department of Commerce, Bureau of the Census, 1990).

23. Joseph M. Dearborn, "The Co-ordination and Administration of City Surveying and Mapping Activities" (paper presented at the Sixth Annual Meeting, American Congress on Surveying and Mapping, Washington, D.C., June 1946). Ronald Abler and John S. Adams, eds., *A Comparative Atlas of America's Great Cities* (Minneapolis: Association of American Geographers and University of Minnesota Press, 1976), treats cartographically (and with uniform scale) twenty metropolitan areas in the United States where most of the nation's people live. The topics covered include place, housing, and population characteristics, providing an excellent overview at the time of publication.

24. Francis A. Walker, *Statistical Atlas of the United States* (New York: J. Bien, 1874). This work, based on the 1870 census, was a pioneering general atlas but was not followed up until the publication of Arch C. Gerlach, ed., *National Atlas of the United States* (Washington, D.C.: United States Geological Survey, 1970). *The Atlas of Britain* (Oxford: Clarendon Press, 1963), planned and directed by D. P. Bickmore and M. A. Shaw, lists approximately fifty national atlases (vii), to which have been added many more since the publication of that work.

25. Le Gear, "Atlases: A List."

26. J. D. Chapman and D. B. Turner, eds., and A. L. Farley and R. I. Ruggles, cartographic eds., *British Columbia Atlas of Resources* (Vancouver: British Columbia Natural Resources Conference, 1956); and William G. Dean, ed., and G. J. Matthews, cartographer, *Economic Atlas of Ontario* (Toronto: University of Toronto Press and the Government of Ontario, 1969). From the same source as the latter is the three-volume *Historical Atlas of Canada,* a splendid cartographic achievement involving many scholars over a number of years and supported by the Social Sciences and Humanities Research Council of Canada.

27. Zink (Tyner), "Lunar Cartography," and "Early Lunar Cartography." Also W. F. Ryan, "John Russell, R.A., and Early Lunar Mapping," *The Smithsonian Journal of History* 1 (1966): 27–48. Russell (1745–1806), a prominent artist, produced excellent shaded-relief renderings of the moon (for which he gathered data by serious astronomical observation) in pastel and other media. In his time there was no adequate means of reproducing these continuous tonal drawings directly; indeed, before the development of modern reproduction techniques, artists, like cartographers, had to rely on the interpretations of engravers for

multiple copies of their work—unless, of course, they did their own engraving, as was true in some cases. Before the invention of the telescope, William Gilbert (1544–1603) in England attempted to delineate the relief of the moon; his fellow countryman Hariot drew maps of the moon and other celestial bodies with 6 to 30X telescopes between 1609 and 1613.

28. Leif Anderson and E. A. Whitaker, *NASA Catalogue of Lunar Nomenclature,* NASA Reference Publication 1097 (Washington, D.C.: NASA, 1982).

29. The International Astronomical Union (IAU), founded in 1919, established rules and conventions for naming features on planets and their satellites. *Toponymy* (geographical nomenclature) in the United States is officially the concern of the Board on Geographic Names, U.S. Department of the Interior.

*Chapter Nine*

1. It has been remarked that the slavish copying of Newtonian calculus notations in the 1700s set English mathematics back a century, during which progress was being made on the Continent. The same might be said of the artistic influence in Italy of Michelangelo (1475–1564) in the century following his death.

2. Michael Ward, "The Mapping of Everest," *The Map Collector* 64 (autumn 1993): 10–15, with illustrations and bibliography.

3. Sir Herbert George Fordham, *Studies in Carto-bibliography* (Oxford: Clarendon Press, 1914; reprint, London: Dawson, 1969) discusses principles of cartobibliography, a term coined by Fordham and followed, with variations, by librarians, archivists, collectors, and others to the present. Eric Wolf, *The History of Cartography: A Bibliography, 1981–1992* (Falls Church, Va.: The Washington Map Society and FIAT LUX, 1992). For other examples, see Black, T. Campbell, Cumming, Ehrenberg, Karrow, Le Gear, Nebenzahl, Ristow, Shirley, Skelton, Stephenson, Tyacke, Verner, and Wallis (all previously cited) and publications by Roy V. Boswell, Philip Lee Phillips, and John Wolter in English, and equal numbers in other languages. Through "metadata," the computer is providing new avenues for accessing sources used to compile maps.

4. *National Geographic Index, 1947–1983* (Washington, D.C.: National Geographic Society, 1983), plus annual lists since 1983, and *Index to Place Names,* published by the NGS previously with each map; look under the entry "map supplements" in the indexes for these contributions. We have previously mentioned (chap. 8) "Space Portrait U.S.A." utilizing NASA (Landsat) imagery and published by the NGS, a mix of private and government resources. The society also published *Historical Atlas of the United States* (1988) with much innovative cartography, graphics, photography, and a substantial text covering a wide variety of topics.

5. Jacques May, "Medical Geography: Its Methods and Objectives," *The Geographical Review* 40 (1950): 9–41; one example in a series by the same author is "Map of the World Distribution of Cholera," *The Geographical Review* 41 (1951): 272–73, plus inset map. Also J. K. Wright, "Cartographic Considerations: A Proposed Atlas of Diseases," *The Geographical Review* 34 (1944): 649–52. For more recent work in this field, see Andrew D. Cliff and Peter Haggett, *Atlas of Disease*

*Distribution: Analytic Approaches to Epidemiological Data* (Oxford: Blackwell, 1988), and one of many sheet maps, "AIDS in LA," drawn under the direction of William Bowen by students at California State University, Northridge, 1988.

6. Executive Committee, Association of American Geographers, "Map Supplement to the *Annals*," *Annals of the Association of American Geographers* 48, no. 1 (March 1958): 91.

7. W. William-Olsson, "Report of the I.G.U. Commission on a World Population Map," *Geografiska Annaler* 45, no. 4 (1963): 243–91, consists of a series of articles on different aspects of population mapping. The map is titled "California Population Distribution in 1960" by Norman J. W. Thrower, *Annals of the Association of American Geographers* 56, no. 2 (1966).

8. Gosta Ekman, Ralf Lindman, and W. William-Olsson, "A Psychophysical Study of Cartographic Symbols," in William-Olsson, "Report": 262–71.

9. The most general and important work on the subject of nonverbal symbols is Henry Dreyfuss, *Symbol Sourcebook: An Authoritative Guide to International Graphic Symbols* (New York: McGraw-Hill Book Company, 1972), by a leading industrial engineer, with a foreword by R. Buckminster Fuller and a good bibliography. See also Edward R. Tufte, *The Visual Display of Quantitative Information* (Cheshire, Conn.: Graphics Press, 1983). Earlier studies on individual symbols used in thematic cartography include Robert L. Williams, "Equal-Appearing Intervals for Printed Screens," *Annals of the Association of American Geographers* 48 (1958): 132–39; George F. Jenks and Duane S. Knos, "The Use of Shaded Patterns in Graded Series," *Annals of the Association of American Geographers* 51 (1961): 316–34; J. Ross MacKay, "Dotting the Dot Map," *Surveying and Mapping* 9 (1949): 3–10; Richard E. Dahlberg, "Towards the Improvement of the Dot Map," *The International Yearbook of Cartography* 7 (1967): 157–66; J. Ross MacKay, "Some Problems and Techniques of Isopleth Mapping," *Economic Geography* 27 (1951): 1–9; Philip W. Porter, "Putting the Isopleth in Its Place," *Proceedings of the Minnesota Academy of Science* 25–26 (1957–58): 372–84; James J. Flannery, "The Graduated Circle: A Description, Analysis and Evaluation of a Quantitative Map Symbol" (Ph.D. diss., University of Wisconsin, 1956); and Alan M. MacEachren and D. R. Fraser Taylor, eds., *Visualization in Modern Cartography* (Oxford: Pergamon, 1994). Also, several major cartographic elements and symbols are considered in Arthur H. Robinson, *The Look of Maps: An Examination of Cartographic Design* (Madison: University of Wisconsin Press, 1952); Norman J. W. Thrower, "Relationship and Discordancy in Cartography," *The International Yearbook of Cartography* 6 (1966): 13–24; Wallis and Robinson, *Cartographical Innovations;* and Jacques Bertin, *Semiologie Geographique* (Paris: Mouton, 1967), translated by William J. Berg and published in English as *Semiology of Graphics: Diagrams, Networks, Maps* (Madison: University of Wisconsin Press, 1983).

10. A useful general classification of map symbols was made by John K. Wright, who categorized point, line, surface, and volume symbols and further subdivided these; see "Atlas of Diseases."

11. Books on map production include Anson and Ormeling, *Basic Cartography* and, for procedures at an earlier period, J. S. Keates, *Cartographic Design and Production* (London: Longman Group Limited, 1978), and A. G. Hodgkiss, *Maps*

*for Books and Theses* (New York: Pica Press, 1970). Although cartographic techniques change rapidly, the principles of design are more enduring, as illustrated by these books and D. W. Rhind and D. R. F[raser] Taylor, eds., *Cartography Past, Present, and Future: A Festschrift for F. J. Ormeling* (London and New York: Elsevier, 1989). This is a tribute to the former director of the International Training Center (ITC), an institution founded in Enschede, the Netherlands, for the promotion of education in cartography, with contributions by leading contemporary scholars in the field.

12. Gerald L. Greenberg, "A Cartographic Analysis of Telecommunication Maps" (master's thesis, University of California, Los Angeles, 1963). This work contains a survey and analysis of over forty types of maps used in telecommunications systems and services; they are considered in respect to development, sources, design, scale, projection, and so forth. Among the maps discussed and illustrated in this work are radio, telephone, communications, Loran, radar, network, and engineering charts.

13. The London Transport map, "The London Underground," designed by Paul E. Garbutt, is considered a masterpiece of such cartography and has served as a model for other rapid-transit diagrams.

14. James R. Akerman, "Blazing A Well-worn Path: Cartographic Commercialism, Highway Promotion, and Automobile Tourism in the United States, 1880–1930," *Cartographica* 44 (1993), with an excellent bibliography on the subject. Otto G. Lindberg, *My Story* (Convent Station, N.J.: General Drafting Co., Inc., 1955), is an "Horatio Alger" tale of the American automobile map business by the former CEO of one of the "big three" producers of these products in the United States. See also Roderick C. McKenzie, "The Development of Automobile Road Guides in the United States" (M.A. thesis, University of California, Los Angeles, 1963).

15. Kitirô Tanaka, "The Relief Contour Method of Representing Topography on Maps," *The Geographical Review* 40 (1950): 444–56.

16. *Technical Report EP-79: Environmental Handbook for the Camp Hale and Pike's Peak Areas, Colorado* (Natick, Mass.: Quartermaster Research and Engineering Center, Environmental Protection Research Division, 1958). Figure 2 of the above report is the map printed in color; figure 11 has an overprint of surface types and figure 14 of vegetation types. See also Arthur H. Robinson and Norman J. W. Thrower, "A New Method of Terrain Representation," *Geographical Review* 47 (1957): 507–20; "On Surface Representation Using Traces of Parallel Inclined Planes," *Annals of the Association of American Geographers* 59 (1969): 600–604; and Norman J. W. Thrower, "Extended Uses of the Method of Orthogonal Mapping of Traces of Parallel, Inclined Planes with a Surface, Especially Relief," *International Yearbook of Cartography* 3 (1962): 26–39, with figure 9.5 from the current text printed in color, following 32.

17. Richard E. Harrison, *Look at the World: The Fortune Atlas for World Strategy* (New York: Knopf, 1944). This is an atlas of perspective-type terrain renderings, many on orthographic projections, showing the globe from unusual orientations, and it is based on Harrison's work for *Fortune* magazine during World War II. Later Harrison usually drew shaded relief in plan view; the most readily avail-

able source of this form of Harrison's renderings is found in Thrower, *Man's Domain*. Figure 9.6 is from this source.

18. Eduard Imhof, *Schweizerischer Mittelschulatas* (Zürich: Konferenz der Kantonalen Erziehungsdirektoren, 1963). This popular European atlas was the practical result of the author's lifelong concern with terrain representation, the theoretical basis of which is contained in his *Kartographische Geländedarstellung* (Berlin: Walter de Gruyter and Co., 1965). John S. Keates, "The Perception of Colour in Cartography," *Proceedings of the Cartographic Symposium* (Edinburgh, 1962): 19–28.

19. See Cholley. Later derivatives of this atlas do not contain line analglyphs and special spectacles for viewing these.

20. George F. Jenks and Fred C. Caspall, *Vertical Exaggeration in Three-Dimensional Mapping*, Technical Report 2, NR 389-146, and George F. Jenks and Paul V. Crawford, *Viewing Points for Three-Dimensional Maps*, Technical Report 3, NR 389-146 (Lawrence: Department of Geography, University of Kansas, 1967); George F. Jenks and Michael R. C. Coulson, "Class Intervals for Statistical Maps," *The International Yearbook of Cartography* 3 (1963): 119–34. Since the 1970s, former graduate students of Jenks—notably Jean Claude Muller, Patricia Gilmartin, Terry Slocum, and Robert McMaster—have continued and expanded the quantitative cartographic interests of their mentor.

21. John C. Sherman and Willis R. Heath, "Problems in Design and Production of Maps for the Blind," *Second International Cartographic Conference*, series 2, no. 3 (1959): 52–59. "Large Print Map of Metropolitan Washington, D.C." and "Key" to the above, design and production directed by John C. Sherman (Washington, D.C.: Department of the Interior, Defense Mapping Agency, 1976). Although published by a U.S. government agency, the map and key (in Braille) were inspired by research conducted at the University of Washington by Sherman and his graduate students, some of whom have carried on this line of research.

22. Arthur H. Robinson, "The Cartographic Representation of the Statistical Surface," *The International Yearbook of Cartography* 1 (1961): 53–63; and Thrower, "Extended Uses."

23. Gillian Hill, *Cartographical Curiosities* (London: The British Library, British Museum Publications Ltd., 1978). This work is based on an exhibit with the same title mounted by the British Library in April 1978 and is an ongoing feature in *The Map Collector*.

24. Mark Monmonier, *Maps with the News: The Development of American Journalistic Cartography* (Chicago: University of Chicago Press, 1979), an excellent treatment of the subject, well documented and illustrated with a good bibliography; Walter W. Ristow, "Journalistic Cartography," *Surveying and Mapping* 17, no. 4 (1957): 369–90; James F. Horrabin, *An Atlas of Current Affairs* (London: Victor Gollancz Ltd., 1934). The maps in this last atlas are fine examples of simple journalistic political maps from the period between the wars by a master of this cartographic form. See also Patricia A. Caldwell, "Television News Maps: An Examination of Their Utilization, Content and Design" (Ph.D. diss., University of California, Los Angeles, 1979), and "Television News Maps and Desert Storm,"

*Bulletin of the American Congress on Surveying and Mapping* 133 (August 1991): 30–33.

25. Judith Tyner, "Persuasive Cartography: An Examination of the Map as a Subjective Tool of Communication" (Ph.D. diss., University of California, Los Angeles, 1974), and "Persuasive Cartography," *Journal of Geography* 81, no. 4 (1982): 140–44. Mark Monmonier, *How to Lie with Maps* (Chicago: University of Chicago Press, 1991); Louis O. Quam, "The Use of Maps in Propaganda," *The Journal of Geography* 42, no. 1 (1943): 21–32. Aleksandr S. Sudakov, "A Bit off the Map," *Unesco Courier* (June 1991): 39–40, describes and illustrates how maps of "low information value" were available to average citizens of the USSR and to tourists, while much more accurate representations were used by officials.

26. Peter Gould and Rodney White, *Mental Maps* (Harmondsworth, England: Penguin Books, 1974); the map is reproduced in this book (38) along with other maps of the mind. See also J. Russell Smith and M. Ogden Phillips, *North America* (New York: Harcourt, Brace and Company, 1949), 169; the map by Daniel K. Wallingford was earlier reproduced (169) in this well-known geography text. It was republished, along with other maps in which cartographic geometry has been modified, in Waldo R. Tobler, "Geographic Area and Map Projections," *The Geographical Review* 53 (1963): 59–78. See also John E. Dornbach, "The Mental Map," *Annals of the Association of American Geographers* 49 (1959): 179–80 (abstract).

27. Pierre George, *Introduction a l'etude géographique de la population du monde* (Paris: L'Institut National d'Études Demographiques, 1951). See also William Bunge, *Theoretical Geography* (Lund, Sweden: Lund Studies in Geography, 1966). Cartograms are one of many map forms illustrated and discussed critically in this work.

28. László Lackó, "The Form and Contents of Economic Maps," *Tijdschrift Voor Econ. En Soc. Geografie* 58 (1967): 324–30.

29. Bunge, *Theoretical Geography,* 54; Tobler, "Geographic Area": 65.

30. Jean Dollfus, *Atlas of Western Europe* (Chicago: Rand McNally and Company, 1963).

31. *The International Atlas* (Chicago: Rand McNally and Company, 1969), and, from the same publisher, *The New International Atlas,* anniversary edition (1991). A fully computerized general, commercial atlas, *Hammond Atlas of the World* (Maplewood, N.J.: Hammond Incorporated, 1992), is now available. The number of general, private atlases is large, an outstanding example being *The Times Atlas of the World: Eighth Comprehensive Edition* (New York: Random House, 1990), from the well-known cartographic establishment of Bartholomew, Edinburgh, Scotland.

32. Norman J. W. Thrower, "The City Map and Guide Information," *Surveying and Mapping* 22, no. 4 (1962): 597–98.

33. Most map projections are originally prepared from tables, but they may be transformed from one scale or form to another optically, graphically, or— now commonly, and most expeditiously—by computer.

34. Only certain shapes are likely to be useful. In addition to the regular figures tangential to the globe (tetra-, hexa-, octa-, dodeca-, and icosahedron),

other forms have been used. These include toroidal (doughnut shape), parabo-
loidal, hyperboloidal, pyramidal, and cataclysmal!

35. Max Eckert, *Die Kartenwissenschaft*, 2 vols. (Berlin and Leipzig: Walter
de Gruyter, 1921–25), was the most comprehensive general study of cartography
up to its time. See also Arthur H. Robinson and Barbara Bartz Petchenik, *The
Nature of Maps: Essays toward Understanding Maps and Mapping* (Chicago: University of Chicago Press, 1976), for further thoughts on cartographic theory.

36. Robert B. McMaster and Norman J. W. Thrower, "The Early Years of
American Academic Cartography: 1920–45," *Cartography and Geographic Information Systems* 18, no. 3 (1991): 154–55, on the importance of Goode's contributions. Most of this issue was devoted to a consideration of the work of Goode,
Raisz, Guy-Harold Smith, and the "big three" American cartographic educators
of the period 1950–80—Robinson, Jenks, and Sherman—and their graduate
students. This situation has changed in the post-1980 period with the establishment of a U.S. National Center for Geographic Information and Analysis with
several branches, an outgrowth of Sherman's work.

37. Richard E. Dahlberg, "Maps without Projections," *Journal of Geography*
60 (1961): 213–18, and "Evolution of Interrupted Map Projections," *The International Yearbook of Cartography* 2 (1962): 36–54.

38. Fisher and Miller, *World Maps and Globes*, esp. 79; and Osborn M. Miller,
"Notes on Cylindrical World Map Projections," *The Geographical Review* 32
(1942): 424–30.

39. Tau Rho Alpha, Scott W. Starratt, and Cecily C. Chang, "Make Your
Own Earth and Tectonic Globes," Open-File Report 93-380-A (U.S. Department
of the Interior, U.S. Geological Survey). This outreach project of the USGS is
aimed particularly at children. The professional journal concerned with geographical education in the United States is the *Journal of Geography*, published
under different titles since 1897 and containing occasional articles on maps and
globes. See Anne Geissman Canright, "Elementary Schoolbook Cartographics:
The Creation, Use, and Status of Social Studies Textbook Maps" (Ph.D. diss.,
University of California, Los Angeles, 1987), for the cartographic "fare" offered
to American children, most of whom never see an atlas.

40. Snyder, *Flattening the Earth*, 188–89.

41. Harry P. Bailey, "A Grid Formed of Meridians and Parallels for the Comparison and Measurement of Area," *The Geographical Review* 46, no. 2 (1956):
239–45, and "Two Grid Systems That Divide the Entire Surface of the Earth into
Quadrilaterals of Equal Area," *Transactions of the American Geophysical Union* 37,
no. 5 (October 1956): 628–35.

42. To illustrate how profound this change has been, the leading U.S. professional journal in the field was renamed in 1990 from *The American Cartographer*
to *Cartography and Geographic Information Systems;* at about the same time, the
then-newly-appointed director of Britain's Ordnance Survey was an academic
GIS specialist after two centuries of military or quasi-military control. See J. C.
Muller, ed., *Advances in Cartography* (London and New York: Elsevier for the International Cartographic Association, 1991), a series of essays on new directions
in cartography/GIS by thirteen practitioners/educators covering data bases, de-

sign, and so on. For an overview of the short history of GIS—from punch cards and mainframe computers in the late 1950s to the use of minicomputers, work stations, and personal computers in the 1990s—see J. T. Coppock and D. W. Rhind, "The History of GIS," in *Geographical Information Systems,* ed. David J. Maguire, Michael F. Goodchild, and David W. Rhind, 2 vols. (New York: Longman, 1991), 21–43.

43. The literature on GIS, though recent, is voluminous and includes P. A. Burrough, *Principles of Geographical Information Systems for Land Resource Assessment* (Oxford: Clarendon Press, 1986); James D. Carter, *Computer Mapping: Progress in the Eighties,* Resource Publications in Geography (Washington, D.C.: Association of American Geographers, 1984); Mark Monmonier, *Technological Transition in Cartography* (Madison: University of Wisconsin Press, 1985); David Rhind and Tim Adams, *Computers in Cartography* (London: British Cartographic Society, 1982); D. R. Fraser Taylor, ed., *The Computer in Contemporary Cartography* (Chichester, England: John Wiley and Sons, 1980); Graeme F. Bonham-Carter, *Geographic Information Systems for Geoscientists* (Oxford: Pergamon, forthcoming); and, on a more specialized but fundamental topic, Robert B. McMaster and K. Stuart Shea, *Generalization in Digital Cartography* (Washington, D.C.: Association of American Geographers, 1992).

44. Joan Baum, *The Calculating Passion of Ada Byron* (Hamden, Conn.: Archon Books, 1986). Lady Lovelace deserves to be a major icon of the women's movement, but her contribution is hardly noticed in the general literature. As indicated by citations in this study, women have made outstanding contributions to cartography in the modern period, and their numbers in training programs are now about equal to those of men.

45. For earlier uses of the computer in mapping, see Waldo R. Tobler, "Automation and Cartography," *The Geographical Review* 49 (1959): 526–34; and William Warntz, "A New Map of the Surface of Population Potentials for the United States, 1960," *The Geographical Review* 54 (1964): 170–84. The SYMAP Project was administered through the Laboratory for Computer Graphics of Harvard University, while an English method, the Oxford Mark I System, was described in W. G. V. Balchin and Alice M. Coleman, "Cartography and Computers," *The Cartographer* 4, no. 2 (1967): 120–27. For one of many thematic atlases that are largely computer-generated, see James Paul Allen and Eugene James Turner, *We the People: An Atlas of America's Ethnic Diversity* (New York: Macmillan, 1988), with well-designed graphics.

46. The map "Northridge Earthquake Epicenters as of January 24, 1994," was published by Environmental Systems Research Institute (ESRI), Redlands, California, and from the same source, *ARC/INFO MAPS* (1988), containing many examples. Matthew McGrath, "Methods of Terrain Representation Using Digital Elevation Data," (M.A. thesis, University of California, Los Angeles, 1990).

47. Norman J. W. Thrower, "Animated Cartography," *The Professional Geographer* 11, no. 6 (1959): 9–12; "Animated Cartography in the United States," *International Yearbook of Cartography* 1 (1961): 20–30; see also Waldo R. Tobler, "A Computer Movie Simulating Urban Growth in the Detroit Region," *Economic Geography* 46, no. 2 (1970): 234–40. Following these three articles there was a hiatus

in interest in the subject, but recently there have been a number of studies on film animation, such as Alan M. MacEachren and David Di Biase, "Animated Maps and Aggregate Data: Conceptual and Practical Problems," *Cartography and Geographic Information Systems* 18, no. 4 (1991): 221–29; and Christopher R. Weber and Barbara P. Buttenfield, "A Cartographic Animation of Average Yearly Surface Temperature for the Forty-eight Contiguous United States, 1897–1986, *Cartography and Geographic Information Systems* 20, no. 3 (1993): 141–50, with ample bibliographies. On seeing the first articles about animated cartography, R. A. Skelton, then superintendent of the Map Room at the British Museum, remarked that this development would require new viewing equipment for such institutions, which has, in fact, now been installed.

48. H. Clifford Darby, "Historical Geography," in *Approaches to History*, ed. H. P. R. Finberg (London: Routledge and Kegan Paul, 1962), 127–56, esp. 139. In commenting favorably on the potential of animated cartography for studies in historical geography, Professor Darby pointed out the problems of obtaining data and the lack of explanation. The latter objection can now be overcome by "scrolling in" a text and by other techniques of interactive video.

49. Norman J. W. Thrower and Helen M. Wallis, "Columbus: The Face of the Earth in the Age of Discovery," a package available from the UCLA Center for Medieval and Renaissance Studies, including the Micro Floppy Disk Columbus Docs MacWrite, Documentation for *Columbus' First Voyage* (HyperCard version by Lisa L. Spangenberg), and *Diario* Transcription and Translation.

50. Such a system, with a playing time of over one hundred hours, was produced by filmmaker Robert Abel through his company, Synapse Corporation of Los Angeles, and IBM for the Columbus Quincentenary, for which the author served as a consultant.

*Appendix B*

1. John K. Wright, "The Terminology of Certain Map Symbols," *Geographical Review* 34 (1944): 653–54.

2. Horn, "Die Geschichte der Isarithmenkarten"; J. L. M. Gulley and K. A. Sinnhuber, "Isokartographie: Eine Terminological Studie," *Kartographische Nachrichten* 4 (1961): 89–99; The Royal Society, *Glossary of Technical Terms in Cartography* (London: The Royal Society, 1966).

# Illustration Sources

Cover illustration: part of Halley's Atlantic isogonic chart, 1701, in the "first state" (without dedication) from William Andrews Clark Memorial Library and Center for Seventeenth- and Eighteenth-Century Studies, University of California, Los Angeles, color negative courtesy of Rand McNally and Company; and part of the map, "Mare Nectaris—Mare Imbrium" from U. S. Defense Mapping Agency. Both reprinted with permission.

*Fig. 1.1:* Walter Blumer, "The Oldest Known Plan of an Inhabited Site Dating from the Bronze Age," *Imago Mundi* 18 (1964): between 8 and 9. Reprinted with permission of the editor of Imago Mundi Ltd. *Fig. 1.2:* Sir Henry Lyons, "The Sailing Charts of the Marshall Islanders," *Geographical Journal* 72 (1928): facing 327. Reprinted with permission of the Royal Geographical Society. *Fig. 1.3:* Bodleian Library, Oxford University. Reproduced by permission. *Fig. 1.4:* G. Malcolm Lewis, "The Indigenous Maps and Mapping of North American Indians," *The Map Collector* 9 (1979): 27. Reprinted with permission of Map Collector Publication, Ltd. *Fig. 1.5:* Deakin University Collection. Reproduced by permission of Mr. Galarrwuy, Yirrkala, NT, for the Gumatj clan.

*Fig. 2.1:* Department of Special Collections, Research Library, UCLA. Prince Youssouf Kamal, *Monumenta cartographica Africae et Aegypti*, tome 1, fascicule 1.6 (1926). Reprinted with permission. *Figs. 2.2 and 2.4:* Eckhard Unger, "Ancient Babylonian Maps and Plans," *Antiquity* 9 (1935): facing 315 and 312. Reprinted with permission of the publisher. *Fig. 2.3:* Theophile J. Meek, "The Orientation of Babylonian Maps," *Antiquity* 10 (1936): plate 8, facing 225. Reprinted by permission of the publisher.

*Figs. 3.1, 3.3, and 3.4:* Joseph Needham and Wang Ling, *Science and Civilisation in China* (Cambridge: Cambridge University Press, 1959), 3: facing 548, facing 552, and fig. 227. Reprinted with permission of the publisher. *Fig. 3.5:* Courtesy George Kish. *Fig. 3.6:* Reginald H. Phillimore, "Three Indian Maps," *Imago Mundi* 9 (1952): between 112 and 113. Reprinted with permission of the editor of Imago Mundi Ltd.

*Fig. 4.6:* Cotton MS, Claudius DVI. Reproduced by permission of the British Library. *Figs. 4.7, 4.8, and 4.9:* Department of Special Collections, Research Library, UCLA. Prince Youssouf Kamal, *Monumenta cartographica Africae et Aegypti*, tome 3 fascicule 4 (1934), 864, 858, and 867. Reprinted with permission. *Fig.*

*4.12:* Kenneth Nebenzahl, *Atlas of Columbus and the Great Discoveries* (Chicago: Rand McNally, 1990), 6. Reprinted with permission.

*Fig. 5.1:* The British Library. Reproduced with permission. *Fig. 5.2:* New York Public Library. *Fig. 5.4:* E. G. Ravenstein, *Behaim's Globe* (London: George Philip & Son, Ltd., 1908). Reprinted with permission. *Fig. 5.5:* Norman J. W. Thrower, "New Geographical Horizons: Maps," in *First Images of America: The Impact of the New World on the Old,* ed. Fredi Chiappelli (Berkeley and Los Angeles: University of California Press, 1976). Reprinted with permission of the publisher. *Fig. 5.10:* Department of Special Collections, Research Library, UCLA. Reprinted with permission. *Fig. 5.12:* Department of Special Collections, Research Library, UCLA. Reprinted with permission.

*Fig. 6.2:* Reprinted by permission of the Royal Society, London. *Figs. 6.3 and 6.4:* Reprinted by permission of the Royal Geographical Society, London. *Fig. 6.5:* Sybrandus Johannes, Fockema Andreae, and B. van't Hoff Geschiendeuis der Kartografie van Nederland ('s-Gravenhage: Martinus Nijhoff, 1947), plate 14. Reprinted by permission of the publisher. *Fig. 6.6:* Charles J. Singer, ed., *A History of Technology* (Oxford: Clarendon Press, 1954–58), 4:606. Reprinted by permission of the publisher. *Fig. 6.7:* Add. MS 7085. Reproduced by permission of the British Library. *Figs. 6.8 and 6.9:* Department of Special Collections, Research Library, UCLA. Reprinted by permission. *Fig. 6.10:* Library of Congress. Reproduced with permission.

*Figs. 7.1 and 7.2:* R. A. Skelton, "The Early Map Printer and His Problems, *The Penrose Annual* 57 (1964): 185, 184. Reprinted by permission. *Figs. 7.3 and 7.9:* National Archives, Washington, D.C. Courtesy Herman Friis. *Fig. 7.4:* National Archives, Washington, D.C. *Figs. 7.5 and 7.6:* Norman J. W. Thrower, *Original Survey and Land Subdivision: A Comparative Study of the Form and Effect of Contrasting Cadastral Surveys* (Chicago: Association of American Geographers, 1966), 40, 6–7. Reprinted by permission of the publisher. *Fig. 7.7:* Norman J. W. Thrower, "The County Atlas of the United States," *Surveying and Mapping* 21 (1961): 336. Reprinted by permission of the American Congress on Surveying and Mapping, Washington, D.C. *Figs. 7.10 and 7.11:* Arthur H. Robinson, "The 1837 Maps of Henry Drury Harness," *The Geographical Journal* 121 (1955): facing 448, 441. Reprinted by permission of the author and the Royal Geographical Society, London. *Fig. 7.12:* E. W. Gilbert, "Pioneer Maps of Health and Disease in England," *The Geographical Journal* 124 (1958). Reprinted by permission of the author and the Royal Geographical Society, London. *Fig. 7.13:* Norman J. W. Thrower, "Relationship and Discordancy in Cartography," *International Yearbook of Cartography* 6 (1966): 21. Reprinted by permission of the publisher, C. Bertelsmann Verlag. *Fig. 7.14:* Department of Special Collections, Research Library, UCLA. Reprinted by permission. *Fig. 7.15:* William M. Davis, assisted by William H. Snyder, *Physical Geography* (New York: Ginn and Company, 1989), 170.

*Fig. 8.2:* Richard A. Gardiner, "A Re-Appraisal of the International Map of the World (IMW) on the Millionth Scale," *International Yearbook of Cartography* 1 (1961): between 32 and 33. Reprinted by permission of the publisher, C. Bertelsmann Verlag. *Fig. 8.8:* Point Dume Quadrangle by Thomas W. Dibblee Jr. (reduced), Dibblee Geological Foundation, P.O. Box 60560, Santa Barbara, CA

93106. Reproduced with permission. *Fig. 8.9:* L. Dudley Stamp and E. C. Willatts, *The Land Utilisation Survey of Britain: An Outline Description of the First Twelve One-Inch Maps* (London: London School of Economics, 1935), frontispiece. Reprinted by permission of the publisher, Geographical Publication, Berkhamsted, England. *Fig. 8.11: Daily Weather Maps,* Sunday, 15 January 1995. *Fig. 8.12:* "Mare Nectaris–Mare Imbrium," by the U.S. Army Map Service.

    *Fig. 9.1:* Norman J. W. Thrower, "California Population Distribution in 1960," *Annals of the Association of American Geographers* 56, no. 2 (June 1966). Reprinted by permission of the publisher. *Fig. 9.4:* Arthur H. Robinson and Norman J. W. Thrower, "A New Method of Terrain Representation," *The Geographical Review* 47 (1957). Reprinted by permission of the publisher. *Fig. 9.5:* Norman J. W. Thrower, "Extended Uses of the Method of Orthogonal Mapping of Traces of Parallel, Inclined Planes with a Surface, Especially Terrain," *International Yearbook of Cartography* 3 (1963): fig. 4, between 32 and 33. Reprinted by permission of the publisher, C. Bertelsmann Verlag. *Fig. 9.7:* Hermann Bollmann, "New York Picture Map," Pictorial Maps, Warren Street, New York City. Reprinted by permission of the publisher. *Fig. 9.8:* Reproduced by permission of Paul Hughes. *Fig. 9.9:* Adapted from Lászlo Lackó, "The Form and Contents of Economic Maps," *Tijdschrift Voor Econ. En Soc. Geografie* 58 (1967): 327–28. *Fig. 9.12:* University of California, Los Angeles, Campus Computing Network (UCLA-CCN). Reprinted with permission. *Fig. 9.13:* Graphic image provided courtesy of Environmental Systems Research Institute, Inc., and California Institute of Technology Seismological Laboratory, Pasadena. Portions of this image are copyrighted by Thomas Bros. Maps and reproduced with permission granted by Thomas Bros. Maps are used by permission of Jack Dangermond. *Fig. 9.14:* Courtesy of the UCLA Center for Medieval and Renaissance Studies.

# Index

Numbers in boldface refer to pages with illustrations.

301